The Practical Approach Series

★ indicates new and forthcoming titles

Affinity Chromatography
Anaerobic Microbiology
Animal Cell Culture (2nd edition)
Animal Virus Pathogenesis
Antibodies I and II
★ Antibody Engineering
★ Basic Cell Culture
Behavioural Neuroscience
Biochemical Toxicology
★ Bioenergetics
Biological Data Analysis
Biological Membranes
Biomechanics—Materials
Biomechanics—Structures and Systems
Biosensors
★ Carbohydrate Analysis (2nd edition)
Cell–Cell Interactions
The Cell Cycle
★ Cell Growth and Apoptosis
Cellular Calcium
Cellular Interactions in Development
Cellular Neurobiology
Clinical Immunology
Crystallization of Nucleic Acids and Proteins
★ Cytokines (2nd edition)
The Cytoskeleton
Diagnostic Molecular Pathology I and II
Directed Mutagenesis
★ DNA Cloning 1: Core Techniques (2nd edition)
★ DNA Cloning 2: Expression Systems (2nd edition)
★ DNA Cloning 3: Complex Genomes (2nd edition)
★ DNA Cloning 4: Mammalian Systems (2nd edition)
Electron Microscopy in Biology
Electron Microscopy in Molecular Biology
Electrophysiology
Enzyme Assays

★ Epithelial Cell Culture
Essential Developmental Biology
Essential Molecular Biology I and II
Experimental Neuroanatomy
★ Extracellular Matrix
Flow Cytometry (2nd edition)
★ Free Radicals
Gas Chromatography
Gel Electrophoresis of Nucleic Acids (2nd edition)
Gel Electrophoresis of Proteins (2nd edition)
★ Gene Probes 1 and 2
Gene Targeting
Gene Transcription
Glycobiology
Growth Factors
Haemopoiesis
Histocompatibility Testing
★ HIV Volumes 1 and 2
Human Cytogenetics I and II (2nd edition)
Human Genetic Disease Analysis
Immunocytochemistry
In Situ Hybridization
Iodinated Density Gradient Media
★ Ion Channels
Lipid Analysis
Lipid Modification of Proteins
Lipoprotein Analysis
Liposomes
Mammalian Cell Biotechnology
Medical Bacteriology
Medical Mycology
★ Medical Parasitology
★ Medical Virology
Microcomputers in Biology
Molecular Genetic Analysis of Populations
Molecular Genetics of Yeast
Molecular Imaging in Neuroscience
Molecular Neurobiology
Molecular Plant Pathology I and II
Molecular Virology
Monitoring Neuronal Activity
Mutagenicity Testing
★ Neural Cell Culture
Neural Transplantation
Neurochemistry
Neuronal Cell Lines
NMR of Biological Macromolecules
★ Non-isotopic Methods in Molecular Biology
Nucleic Acid Hybridization
Nucleic Acid and Protein Sequence Analysis
Oligonucleotides and Analogues
Oligonucleotide Synthesis
PCR 1
★ PCR 2
★ Peptide Antigens
Photosynthesis: Energy Transduction
★ Plant Cell Biology
★ Plant Cell Culture (2nd edition)

Plant Molecular Biology
Plasmids (2nd edition)
★ Platelets
Pollination Ecology
Postimplantation Mammalian Embryos
Preparative Centrifugation
Prostaglandins and Related Substances
Protein Blotting
Protein Engineering
Protein Function
Protein Phosphorylation
Protein Purification Applications
Protein Purification Methods
Protein Sequencing
Protein Structure
Protein Targeting
Proteolytic Enzymes
★ Pulsed Field Gel Electrophoresis
Radioisotopes in Biology
Receptor Biochemistry
Receptor-Ligand Interactions
RNA Processing I and II
Signal Transduction
Solid Phase Peptide Synthesis
Transcription Factors
Transcription and Translation
Tumour Immunobiology
Virology
Yeast

Epithelial cell culture

A Practical Approach

Edited by

ANDREW J. SHAW

Department of Toxicology
School of Pharmacy
University of London
London WC1N 1AX

at
OXFORD UNIVERSITY PRESS
Oxford New York Tokyo

Oxford University Press, Walton Street, Oxford OX2 6DP

Oxford New York
Athens Auckland Bangkok Bombay
Calcutta Cape Town Dar es Salaam Delhi
Florence Hong Kong Istanbul Karachi
Kuala Lumpur Madras Madrid Melbourne
Mexico City Nairobi Paris Singapore
Taipei Tokyo Toronto

and associated companies in
Berlin Ibadan

Oxford is a trade mark of Oxford University Press

Published in the United States
by Oxford University Press Inc., New York

A catalogue record for this book is available from the British Library

Library of Congress Cataloging in Publication Data
(Data available)

ISBN 0 19 963573 0 (Hbk)
ISBN 0 19 963572 2 (Pbk)

Typeset by Footnote Graphics, Warminster, Wilts
Printed in Great Britain by Information Press, Ltd., Eynsham, Oxon

Preface

In the last two decades, cell culture has progressed from an era when cell maintenance was the primary goal, and we gained a tremendous insight into the basic workings of the cell, to a stage now where cells are cultured in order to create tissue-like structures *in vitro*. These models offer many of the benefits associated with cell and organ cultures and present researchers with the opportunity to investigate the unique morphology, biochemistry, and functions of differentiated cells, and the factors which regulate or perturb these features. Epithelia, endothelia, and mesothelia represent some of the simplest tissues of the body and, not surprisingly, have been amongst the first to be modelled using new cell culture methods. Common amongst these techniques, and central to this book, is the culture of these cell types on porous substrates, a factor which, in part, contributes to an environment conducive to the expression of a differentiated phenotype.

Cell culture models of epithelia and endothelia are increasingly being used in diverse fields of research (including pharmaceutics, pharmacology, toxicology, and cell biology) as valuable new research tools. The aims of this book are:

(a) to provide detailed protocols for modelling different epithelia *in vitro* using primary cultures or cell lines,

(b) to provide detailed protocols for morphologically, biochemically, and functionally characterizing these models in relation to their *in vivo* counterparts, and

(c) to outline the research applications to which these cultures can be applied.

In order to achieve these aims, and to stimulate cross-fertilization between different research fields, an international panel of authors from various scientific backgrounds have contributed to this book. The book is divided into two sections: the first four chapters present the basic principles behind modelling epithelia (Chapter 1) and the techniques used for identifying typical characteristics of these cells, namely the expression of cytokeratins (Chapter 2), cell polarity (Chapter 3), and xenobiotic metabolism (Chapter 4). The remaining chapters describe specific examples of models of endothelia (Chapter 5), absorptive epithelia (Chapter 6), glandular epithelia (Chapters 7 and 8), and protective epithelia (Chapter 9), and detail how they are characterized and applied.

This book assumes that the reader has a basic understanding of the principles of cell culture technique and as such makes no attempt to explain the fundamentals of this subject, e.g. sterilization and cell propagation. Several books in the *Practical approach* series may serve as complementary reading to this book, in particular, *Animal cell culture*, 2nd edition (R. I. Freshney), *Cell growth and division* (R. Baserga), *Cell–cell interactions* (B. R. Stevenson, W. J. Gallin, and D. L. Paul), *The cell cycle* (P. Fantes and R. Brooks), *Electron microscopy in biology* (J. R. Harris), and *Growth factors* (I. McKay and I. Leigh).

Andrew J. Shaw

Contents

List of contributors xv

Abbreviations xvii

1. Modelling epithelial tissues *in vitro* 1

Andrew J. Shaw

1. Epithelia: basic structure, function, and biochemistry 1
Introduction 1
Embryonic origin 4
Tissue regeneration 5
Cytokeratins 6
Polarity 6
Intercellular junctions 7
Basement membrane 9

2. Epithelia: methods of culture 9
Introduction 9
Organ culture 10
Histotypic and organotypic cell culture 11

References 15

2. Identification and localization of cytokeratins 17

Ian C. Mackenzie and Zhirong Gao

1. Introduction 17
Organotypic cultures and epithelial differentiation patterns 17
The cytokeratins 18

2. Immunohistochemical techniques for the demonstration of cytokeratins 19
General considerations 19
Preparation of samples for immunohistochemistry 20
The use of processed tissues for immunohistochemistry 21
Selection of primary antibodies 22
Detection of antibody binding 23
Controls for specificity of staining reactions 24

3. Electrophoresis of cytokeratins 27

4. Western blotting of cytokeratins 32

5. Staining transfers for total protein 33

References 35

3. Assessment of cell polarity 37

Elaine J. Hughson and Robert P. Hirt

1. Introduction 37
Epithelial cell polarity 37
The MDCK cell line as a model system 39

2. Cell culture 40
MDCK cell culture on plastic 41
Culturing epithelial cells on permeable supports 42

3. Assessment of monolayer integrity 45
Assessment of monolayer integrity using horseradish peroxidase 46
Measurement of transepithelial electrical resistance 47
Tightness and polarity 47

4. Morphological methods to assess cell polarity 47
Fluorescence microscopy 48
Electron microscopy 51

5. Biochemical methods 55
Polarized internalization of transferrin 55
Selective cell surface labelling of membrane proteins 56

6. Conclusions 63

Acknowledgements 63

References 63

4. Xenobiotic metabolism in epithelial cell cultures 67

André Guillouzo and Christophe Chesné

1. Introduction 67

2. Hepatocyte cultures 67
Primary hepatocyte cultures 67
Hepatic cell lines 69

3. Preparation of samples 69

4. Determination of drug metabolizing capacity 70

5. Conclusions 83

Acknowledgements 84

References 84

5. Culture and characterization of human endothelial cells 87

Victor W. M. van Hinsbergh and Richard Draijer

1. Introduction 87

2. Materials for the isolation and culture of endothelial cells 87
Sera 87
Preparation of endothelial cell growth factor (ECGF) 88
Coating tissue culture dishes and coverslips 89
Preparation of DiI-acetylated LDL 90
Precautions 91

3. Isolation and culture of human endothelial cells 92
Isolation and culture of umbilical vein endothelial cells 92
Isolation and culture of endothelial cells from adult arteries and veins 94
Isolation of endothelial cells from small pieces of large blood vessels 95
Isolation and culture of microvascular endothelial cells from human foreskin 96

4. Identification of human endothelial cells 99
Morphology, ultrastructure, and von Willebrand factor (vWF) 99
Surface determinants 101
Enzymatic activities and receptors 103
Criteria to determine endothelial cell activation 103

5. Additional culture techniques 103
Storage of endothelial cells in liquid nitrogen 103
Culture of human endothelial cells on porous filters 105
Culture of human endothelial cells on microcarriers 106

6. Perspective 108

Acknowledgement 109

References 109

6. Studying transport processes in absorptive epithelia 111

Per Artursson, Johan Karlsson, Göran Ocklind, and Nicolaas Schipper

1. Introduction 111

2. Culture of Caco-2 cell monolayers 112
Sources of cells 112
Growth and maintenance of Caco-2 cells 112
Assessment of monolayer integrity 114

3. Transport studies 121
Passive transport 122

Active transport 123
Elimination of the unstirred water layer 126
Visualization of transcellular and paracellular permeability 131

4. Technical update 131

References 132

7. A model of glandular epithelium for studying secretion 135

Gérard L. Adessi, Laurent Beck, and Abderrahim Mahfoudi

1. Introduction 135

2. Preparation 136
Cell culture media 136
Animals 137
Matrigel-coated filters 140

3. Isolation and culture of glandular epithelial cells 141
Isolation of glandular epithelial cells 141
Morphology of endometrial cells *in vitro* 143
Subculture of glandular epithelial cells into the bicameral system 143

4. Study of cellular and vectorially secreted proteins by glandular epithelial cells 147

5. Study of the effect of progesterone on proteins secreted into both the apical and basal compartments 149
Hormonal treatment 149
Progesterone receptor content of glandular epithelial cells 150
Effect of progesterone on vectorially-secreted proteins 152

6. Co-culture of glandular epithelial cells with stromal cells in the bicameral system 152

References 156

8. A model of the blood–testis barrier for studying testicular toxicity 159

Anna Steinberger and Andrzej Jakubowiak

1. Introduction 159

2. Two-compartment cultures of Sertoli cells 160
Advantages and potential pitfalls 160
Construction of the culture chambers 162
Sertoli cell isolation 163
Testing for monolayer and tight junction patency 169

3. Effects of testicular toxicants on the blood–testis barrier *in vitro* 170
Cadmium chloride 170
Phthalate esters 174

4. Conclusions 177

Acknowledgements 177

References 177

9. Reconstruction of human skin epidermis *in vitro* 179

Marie-Cécile Lenoir-Viale

1. Introduction 179

2. Techniques used to reconstruct human epidermis *in vitro* 180
The dermis 180
The epidermis 183
Reconstruction of human skin epidermis 187
Culture media 192

3. Methods for studying normal cell differentiation 193
Histology 193
Electron microscopy 194
Immunofluorescence 194
Protein extraction and identification 194

4. Uses of *in vitro* reconstructed human skin epidermis 194
Studying normal epidermal cell differentiation 194
Studying pathological conditions 197
Pharmacological, cosmetic, and pharmaceutical research 197

5. Conclusions 198

Acknowledgements 199

References 199

Appendices

A1 Addresses of suppliers 201

A2 An *in vitro* model of the blood–brain barrier for studying drug transport to the brain 207

A. G. de Boer, H. E. de Vries, P. J. Gaillard, and D. D. Breimer

1. Introduction 207

2. Characterization of BMEC 211
3. Applications of BMEC 211
References 212

Index 215

Contributors

GÉRARD L. ADESSI
Service de Biochimie–Biologie Moléculaire, Centre Hospitalier Universitaire de Besançon, Hôpital Jean Minjoz, Boulevard Fleming, 25030 Besançon cedex, France.

PER ARTURSSON
Department of Pharmaceutics, Uppsala University, Box 580, S-751 23 Uppsala, Sweden.

LAURENT BECK
Service de Biochimie–Biologie Moléculaire, Centre Hospitalier Universitaire de Besançon, Hôpital Jean Minjoz, Boulevard Fleming, 25030 Besançon cedex, France.

D. D. BREIMER
Division of Pharmacology, Leiden/Amsterdam Center for Drug Research, University of Leiden, PO Box 9503, 2300 RA Leiden, The Netherlands.

CHRISTOPHE CHESNÉ
BIOPREDIC SA, 14–18, rue Jean Pecker, Rennes cedex, France.

A. G. DE BOER
Division of Pharmacology, Leiden/Amsterdam Center for Drug Research, University of Leiden, PO Box 9503, 2300 RA Leiden, The Netherlands.

H. E. DE VRIES
Division of Pharmacology, Leiden/Amsterdam Center for Drug Research, University of Leiden, PO Box 9503, 2300 RA Leiden, The Netherlands.

RICHARD DRAIJER
Gaubius Laboratory TNO-PG, PO Box 430, 2300 AK Leiden, The Netherlands.

P. J. GAILLARD
Division of Pharmacology, Leiden/Amsterdam Center for Drug Research, University of Leiden, PO Box 9503, 2300 RA Leiden, The Netherlands.

ZHIRONG GAO
University of Texas, Health Science Center at Houston, Dental Branch/Dental Science Institute, PO Box 20068, Houston TX 77335, USA.

ANDRÉ GUILLOUZO
INSERM, Unité de Recherches Hépatologiques U 49, Hôpital Pontchaillou, 35033 Rennes cedex, France.

ROBERT P. HIRT
Molecular Biology Unit, Department of Zoology, The Natural History Museum, Cromwell Road, London SW7 5BD, UK.

ELAINE J. HUGHSON
Department of Biology, York University, Heslington, York YO1 5YW, UK.

ANDRZEJ JAKUBOWIAK
Department of Obstetrics, Gynecology and Reproductive Sciences, University of Texas Medical School, 6431 Fannin, Suite 3.204, Houston, Texas 77030, USA.

JOHAN KARLSSON
Department of Pharmaceutics, Uppsala University, Box 580, S-751 23 Uppsala, Sweden.

MARIE-CÉCILE LENOIR-VIALE
Centre International de Recherches Dermatologiques GALDERMA, 635, route des Lucioles, 06902 Sophia Antipolis cedex, France.

IAN C. MACKENZIE
University of Michigan School of Dentistry, Ann Arbor, Michigan 48109-1078, USA.

ABDERRAHIM MAHFOUDI
Service de Biochimie–Biologie Moléculaire, Centre Hospitalier Universitaire de Besançon, Hôpital Jean Minjoz, Boulevard Fleming, 25030 Besançon cedex, France.

GÖRAN OCKLIND
Department of Pharmaceutics, Uppsala University, Box 580, S-751 23 Uppsala, Sweden.

NICOLAAS SCHIPPER
Department of Pharmaceutics, Uppsala University, Box 580, S-751 23 Uppsala, Sweden.

ANDREW J. SHAW
Department of Toxicology, School of Pharmacy, University of London, 29–39 Brunswick Square, London WC1N 1AX, UK.

ANNA STEINBERGER
Department of Obstetrics, Gynecology and Reproductive Sciences, University of Texas Medical School, 6431 Fannin, Suite 3.204, Houston, Texas 77030, USA.

VICTOR W. M. VAN HINSBERGH
Gaubius Laboratory TNO-PG, PO Box 430, 2300 AK Leiden, The Netherlands.

Abbreviations

Ab	antibody
ACE	angiotensin-converting enzyme
ACM	astrocyte-conditioned medium
AD	adenine
AP	alkaline phosphatase
ATCC	American Type Culture Collection
ATPase	adenosine 5′-triphosphatase
BBB	blood–brain barrier
BME	Eagle's basal medium
BMEC	brain microvessel endothelial cell
BP	bullous pemphigoid
BSA	bovine serum albumin
Caco-2	a human colon adenocarcinoma-derived cell line
Caps	3-[cyclohexylamino]-1-propanesulfonic acid
CDM	chemically defined medium
CLSM	confocal laser scanning microscopy
CMF-	Ca^{2+} and Mg^{2+} free-
CMRL	Connaught Medical Research Laboratories (media)
CNS	central nervous system
CT	cholera toxin
CVO	circumventricular organ
CYP	cytochrome P450
DAB	3,3-diaminobenzidine
DED	de-epidermized dermis
DEX	dexamethasone
DDSA	dodecenyl succinic anhydride
DiI	dioctadecylindocarbocyanine
DMEM	Dulbecco's modified Eagle's medium
DMF	dimethylformamide
DMP-30	2,4,6-tri[dimethylaminomethyl]phenol
DMSO	dimethyl sulfoxide
DTT	dithiothreitol
E_2	oestradiol
ECACC	European Collection of Animal Cell Cultures
ECGF	endothelial cell growth factor
EDTA	ethylenediamine tetra-acetic acid
EGF	epidermal growth factor
ELISA	enzyme-linked immunosorbent assay
FACS	fluorescence activated cell sorter
FCS	fetal calf serum

FD	fluorescein isothiocyanate-labelled dextrans
FGF	fibroblast growth factor
FITC	fluorescein isothiocyanate
FSH	follicle stimulating hormone
GH	growth hormone
GO	glucose oxidase
GSH	glutathione (reduced form)
GST	glutathione-*S*-transferase
GTP	glutamyl transpeptidase
HBSS	Hanks' balanced salt solution
HC	hydrocortisone
HMM	high molecular mass protein
HRP	horseradish peroxidase
HSA	human serum albumin
Ig	immunoglobulin
IL	interleukin
IN	insulin
IPP	immunoprecipitation
K	keratin
Ki-*ras*	Kirsten *ras* oncogene
LDL	low-density lipoprotein
LH	luteinizing hormone
mAb	monoclonal antibody
MCBD	Molecular, Cellular and Developmental Biology Dept, University of Colorado (media)
MDCK	Madin–Darby canine kidney (a cell line)
MEHP	mono-2-ethylhexyl phthalate
MEM	minimum essential medium
MEP	monoethyl phthalate
MNA	methyl nadic anhydride
MTT	dimethylthiazol diphenyltetrazolium
NBCS	newborn calf serum
NEAA	non-essential amino acids
NEPHGE	non-equilibrium pH gradient electrophoresis
NP-40	Nonidet P-40
PAGE	polyacrylamide gel electrophoresis
P_{app}	apparent permeability coefficient
PBS	phosphate buffered saline
PBSA	phosphate buffered saline without Ca^{2+} and Mg^{2+}
PBSB	phosphate buffered saline with 0.2% (w/v) BSA
P_c	permeability coefficient
PEG	polyethylene glycol
PF	paraformaldehyde
pIg-R	polymeric immunoglobulin receptor

PPD	*p*-phenylenediamine
PRL	prolactin
RLEC	rat liver epithelial cells
RPMI	Roswell Park Memorial Institute (media)
SDS	sodium dodecyl sulfate
SF	serum-free
SSM	serum-supplemented medium
TBS	Tris buffered saline
TCA	trichloroacetic acid
TEM	transmission electron microscopy
TEMED	*N,N,N′,N′*-tetramethylethylenediamine
TER	transepithelial (or transendothelial) electrical resistance
TF	transferrin
TI	tri-iodothyronine
TGF-β	transforming growth factor β
Tris	Tris(hydroxymethyl)-aminomethane
TSH	thyroid stimulating hormone
UDPGA	uridine diphosphate glucuronic acid
u-PA	urokinase-type plasminogen activator
val	valine
vWF	von Willebrand factor
WB	Western blot

1

Modelling epithelial tissues *in vitro*

ANDREW J. SHAW

1. Epithelia: basic structure, function, and biochemistry

1.1 Introduction

The human body is composed of four basic types of tissue: epithelial, connective, muscular, and nervous. Epithelia cover the surfaces, and line the cavities and tubes of the body. These tissues are supported on a basement membrane and consist of close-knit polyhedral cells bound laterally to each other by cell junctions. Their upper (apical) surfaces are free and may either be exposed directly to the air, e.g. epidermis and corneal epithelium, or to a fluid filled space, e.g. intestinal and kidney tubule epithelia. They are invariably avascular tissues which are nourished by the capillaries of the underlying connective tissue. The classification of lining and covering epithelia is based on three morphological characteristics (see *Figure 1* and *Table 1*; refs 1, 2):

(a) the number of cell layers (simple, pseudostratified, or stratified)

(b) the surface cell shape (squamous, cuboidal, or columnar)

(c) the presence of surface specializations (cilia or keratin).

Lining and covering epithelia are critical for protecting the underlying vascularized tissues and for regulating the transport of ions and molecules into and out of these tissues. Most of these epithelia also contain cells specialized for secretion, e.g. the goblet cells of the digestive and respiratory tracts. However, where the need for secretions is out of keeping with, or can not be met by these relatively simple tissues, glands develop. Glands are invaginations of epithelial surfaces formed during embryonic development by proliferation of epithelium into the underlying connective tissue. These glands are referred to as exocrine glands, where they retain a connection (duct) with the surface epithelium, and endocrine glands, where this connection is lost. Exocrine glands, e.g. salivary and sweat glands, are further categorized according to whether they are unbranched (simple) or branched (compound), and by the organization of their secretory portion (tubular and/or acinar). Endocrine

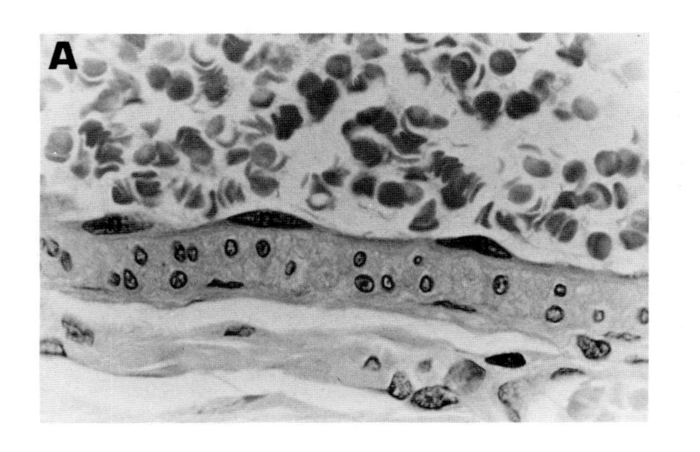
A

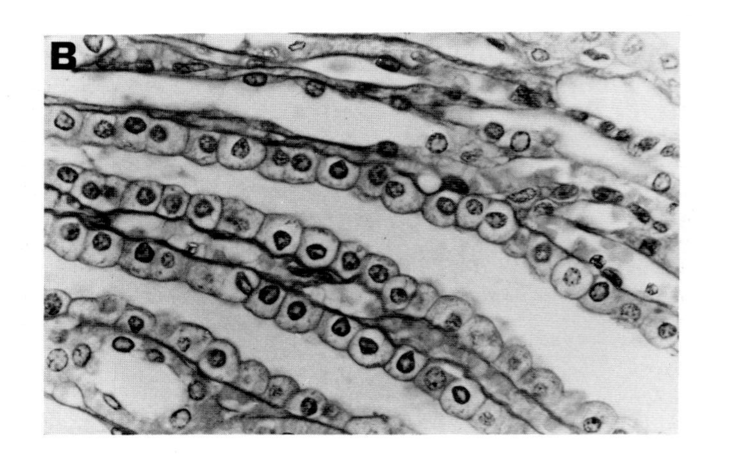
B

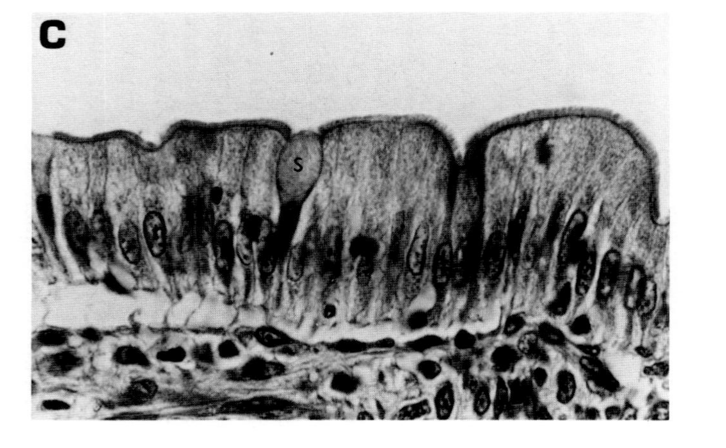
C
S

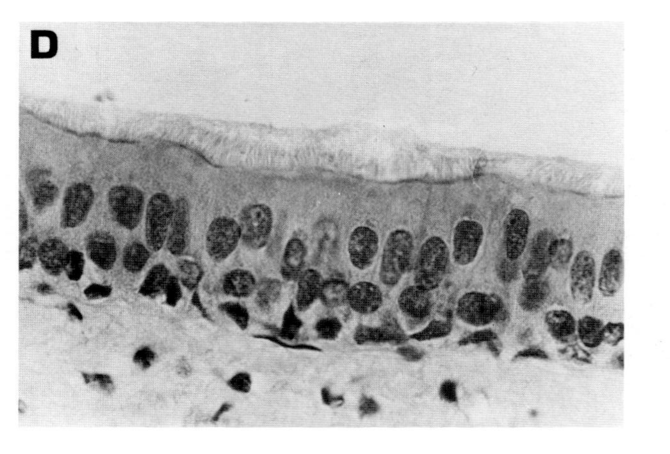
D

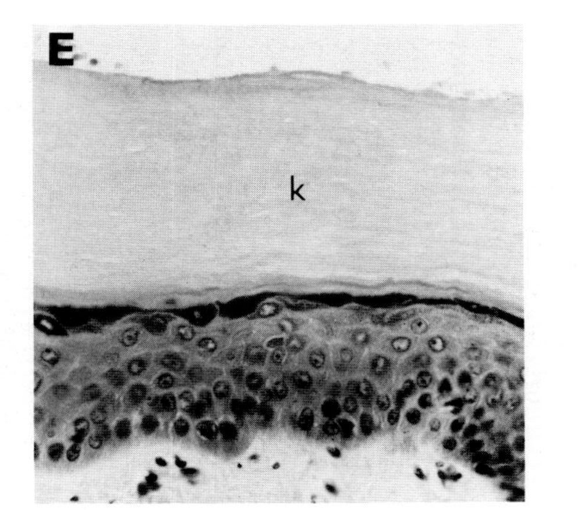

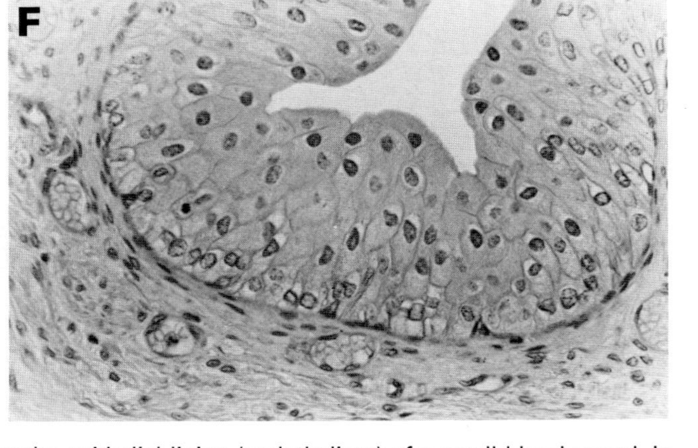

Figure 1. Micrographs of (A) simple squamous epithelium as illustrated by the epithelial lining (endothelium) of a small blood vessel. In this micrograph, three squamous cells can be located by their large elongated nuclei (n) which protrude into the vessel lumen. (B) Simple cuboidal epithelium as illustrated by the lining of the small collecting ducts of the kidney. The nuclei are rounded and sit fairly centrally in the cells. (C) Simple columnar epithelium as illustrated by the lining of the duodenum. The cells are tall and, in this example, the elongated nuclei are located towards the basal membrane, i.e. the cells exhibit polarity (see Section 1.5). The apical (luminal) membrane is characteristic of absorptive epithelia in that it has a brush border comprising numerous microvilli which greatly increases its surface area. A secretory goblet cell (s) is also present. (D) Pseudostratified ciliated epithelium as illustrated by the lining of the trachea. This type of epithelia is distinguishable from simple columnar epithelium because the cells, which all rest on the basement membrane, do not all reach the luminal surface. The cells display polarized features including basally-located nuclei and the presence of cilia on the apical membrane (see Section 1.5). (E) Stratified squamous keratinizing (dry) epithelium as illustrated by the covering (epidermis) of the skin. Stratified epithelia comprise cells undergoing various stages of terminal differentiation, from the cuboidal proliferating basal cells to the squamous, commonly anucleated surface cells. This process of maturation involves the synthesis of different cytokeratins (see Section 1.4) and, in keratinizing epithelium, the formation of a tough keratin barrier (k). This barrier varies in thickness depending on the degree of abrasion that it routinely undergoes, and is especially thick on the sole of the foot as seen in this micrograph. (F) Transitional epithelium as illustrated by the lining of the ureter. In the relaxed state (as shown), the tissue resembles stratified cuboidal epithelium but when stretched is more characteristic of stratified squamous epithelium. All sections and micrographs were kindly prepared by J. A. Turton and D. McCarthy at the School of Pharmacy, University of London.

Table 1. Types of covering and lining epithelia

Number of cell layers	Surface cell shape	Specialization	Examples
Simple (one cell layer)	Squamous		Alveoli, mesothelium (endothelium)
	Cuboidal		Small collecting ducts of kidney, salivary glands, and pancreas
	Columnar		Intestine, stomach, and gall bladder
		Ciliated	Female reproductive tract
Pseudostratified (all cells are attached to the basement membrane but are of different height and shape)		Ciliated	Large airways of the respiratory tract
Stratified (two or more cell layers)	Squamous		Oral cavity, oesophagus, pharynx, anal canal, and vagina
		Keratinized	Epidermis
	Cuboidal		Large excretory ducts of exocrine glands (e.g. sweat glands, salivary glands, and pancreas)
	Transitional		Bladder and ureters
	Columnar		Conjunctiva

glands are characterized as those forming anastomosing cords interspersed between dilated blood capillaries, e.g. adrenal glands, or those which develop follicles, e.g. thyroid glands.

Multicellular glands form organs with an orderly structure and their own blood and nervous systems. These organs may display both exocrine and endocrine functions. For example, in the liver, hepatocytes secrete products directly into both ducts (bile) and the blood (plasma proteins and lipoproteins), while in the pancreas, acinar cells secrete digestive enzymes into the intestinal lumen, and islet cells secrete insulin and glucagon into the blood.

1.2 Embryonic origin

Epithelia are derived from all three germ layers of the embryo. Ectoderm, for example, gives rise to the covering of the skin and body openings, while endoderm generates the lining and glands of the digestive tract, and the lining of the respiratory tract. Epithelia of mesodermal origin are traditionally regarded as not being true epithelia and are classified separately as mesothelium (the serous lining of the pericardial, pleural, and peritoneal cavities) and endothelium (the lining of the blood and lymphatic vessels). The latter in particular differs fundamentally from most true epithelia in rarely expressing cytokeratins (see Section 1.4). However, there are many similarities between these tissues morphologically, functionally and, most importantly with regard to this book, in the methods for remodelling these tissues *in vitro*. Consequently the term *epithelia* is used in this book to refer to all of these tissues.

1.3 Tissue regeneration

The nature of epithelia, especially as a front-line defence against physical, chemical, and microbial insults, requires that the mature (differentiated) cells of these tissues are readily replaced as they are lost. This regeneration is a continuous process, the rate of which is controlled by cell loss. Consequently, where cell trauma is high, e.g. the skin and gastrointestinal tract, regeneration is rapid, while where it is low, e.g. glandular tissues, regeneration is slow. Also, because epithelial cells are responsible for the major functions of many organs, they also need to respond quickly to any sudden cell loss, for example due to injury. Replacement of mature cells generally occurs by one of two routes (see *Table 2* for definition of terminology):

(a) Undifferentiated stem cells produce daughter cells committed to becoming mature cells. These committed precursor cells progressively lose their capacity to proliferate as they develop their characteristic phenotype. This type of regeneration is commonly seen where continual renewal is required, e.g. covering epithelia.

(b) Mature cells dedifferentiate and regain their capacity to proliferate. This type of regeneration is a characteristic response of slowly renewed tissues which have undergone sudden cell loss, e.g. liver regeneration after partial hepatectomy. Endothelial cells can also dedifferentiate, proliferate, and differentiate again during angiogenesis.

Table 2. Definition of cell terminology

Stem cell
An unspecialized, and often slowly dividing cell, which has the potential to progress to one of one or more cell lineages.

Commitment
The progression from a stem cell to a specific cell lineage. This process is irreversible and furnishes the cell with the potential to express certain properties characteristic of that cell lineage.

Differentiation
The development, by a committed cell, of phenotypic properties characteristic of a functionally mature cell. This process may or may not be complete or reversible.

Terminal differentiation
The process leading to the full expression of phenotypic properties by a cell. From this point the cell cannot progress further, and normally cannot dedifferentiate.

Dedifferentiation
The loss of certain phenotypic properties characteristic of the mature cell. This process normally occurs concurrently with an increase in the capacity of the cell to proliferate.

1.4 Cytokeratins

Almost all epithelia (except endothelia) contain intermediate-sized (10 nm) filaments composed of the complex protein, keratin. Twenty different cytokeratins have been identified by two-dimensional gel electrophoresis, and these have been divided into two families (see Chapter 2; refs 3, 4); type I, which contains acidic (pI 4.5–6.0), smaller cytokeratins (40–56.5 kDa), and type II, which contains relatively basic (pI 6.0–8.0), larger cytokeratins (52–67 kDa). Between two and ten different cytokeratins are expressed by each epithelial cell type, characteristically as heterotypic pairs. The expression of cytokeratins in epithelia is governed by three factors:

(a) The type of epithelia: for example, the smallest type I and II cytokeratins are generally restricted to simple epithelia, while the larger cytokeratins are expressed by various stratified epithelia.

(b) The state of differentiation of stratified epithelia: for example, epidermis, corneal epithelium, and oesophageal epithelium express the 58 kDa (K5)/50 kDa (K14) cytokeratin pair in their basal cells, but in their suprabasal layers express the 65 kDa (K1)/56.5 kDa (K10), 64 kDa (K3)/55 kDa (K12), and 59 kDa (K4)/51 kDa (K13) cytokeratin pairs, respectively.

(c) The state of cell growth: for example, under hyperproliferative conditions, all the suprabasal cytokeratins mentioned in (b) diminish and the 56 kDa (K6)/48 kDa (K16) cytokeratin pair is expressed.

1.5 Polarity

An essential feature of all epithelial cells is their polarity. This characteristic is aptly demonstrated by the enterocytes forming the simple columnar epithelium of the intestine (see *Figures 1C* and *2*). The apical surface of these cells is naked to the intestinal lumen while their basal surface is firmly bound to the basement membrane. The cell nuclei are displaced towards the basal surface.

The apical membrane of epithelial cells is morphologically, functionally, and biochemically distinct from the basal and lateral (basolateral) membranes. Absorptive epithelia, for example of the intestine (see *Figure 2* and Chapter 6) and the renal tubule (see Chapter 3), display orderly arrays of microvilli on their apical surface which greatly increase the surface area for absorption. These microvilli are approximately 1 μm high and are covered by a filamentous coat containing glycoproteins (the glycocalyx). Together, the microvilli and glycocalyx make up the brush border of the cell layer. The cells lining the upper respiratory tract possess larger apical structures, cilia (5–10 μm in length), whose coordinated movement is used to generate a unidirectional flow of fluid or particulate matter (the mucociliary escalator).

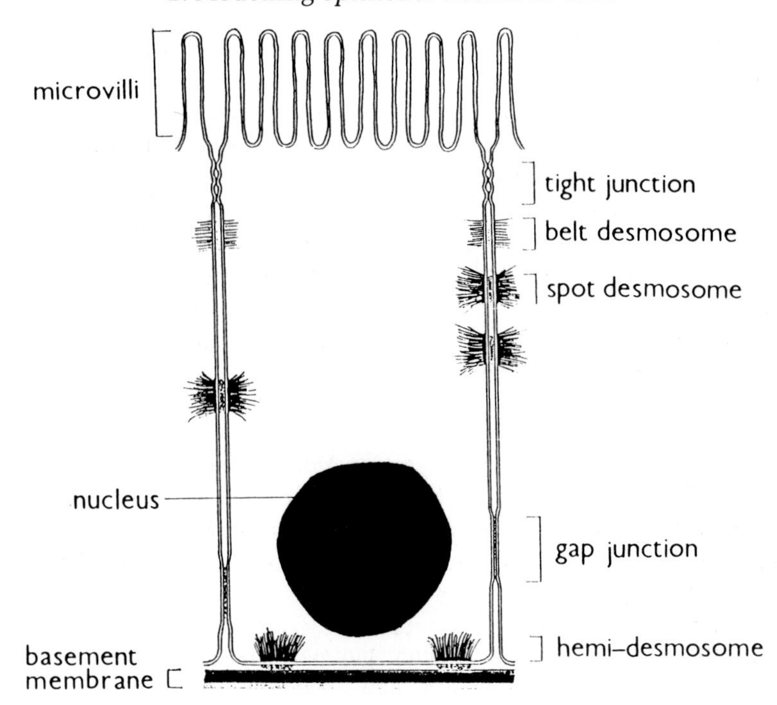

Figure 2. A drawing of an enterocyte emphasizing the polarity of the cell and the various intercellular junctions.

Absorptive epithelia are ideally tailored to regulate the transport of ions and molecules into underlying tissues. The close-knit nature of the cells largely prevents paracellular transport (i.e. movement of materials between the cells) thus allowing the cells to control the movement of chemicals through themselves (transcellular transport) by various means (see Chapter 6). The apical and basolateral membranes are required to fulfil different functions in this process, the former to incorporate molecules and the latter to expel them. Consequently, these membranes possess different proteins, e.g. enzymes and pumps, to fulfil these tasks. The basolateral membrane is also functionally and biochemically distinct because it contains protein pumps for taking up nutrients diffused from the underlying blood vessels, and receptors for neurotransmitters, hormones, and basement membrane components which influence cell behaviour. For the cells to function properly, it is essential that the apical and basolateral membrane proteins remain segregated. This is achieved by the presence of impermeable intercellular junctions.

1.6 Intercellular junctions

A marked characteristic of epithelial tissues, and especially of those subject to abrasive action (e.g. the epidermis), is the strong cohesion between the cells. This is, in part, due to the action of membrane-bound glycoproteins and intercellular proteoglycans: as glycoproteins relinquish their adhesive function in the absence of Ca^{2+} ions, the chelating agent EDTA has been employed for generating single-cell suspensions of epithelial cells. Epithelial cells are also firmly linked to neighbouring cells by intercellular junctions. These are commonly classified into three functional categories:

(a) Adhering junctions (e.g. belt and spot desmosomes), which primarily bind cells together.

(b) Impermeable junctions (e.g. tight junctions), which prevent the flow of material through the intercellular space and thus underpin the regulation of transepithelial transport.

(c) Communicating junctions (e.g. gap junctions), which allow the transfer of small intracellular molecules between adjacent cells. All these junctions appear on the lateral membranes of the cells and, where they are all present, are displayed in a specific order from the apical to the basal surface (see *Figure 2*).

Tight junctions (*zonulae occludentes*; singular, *zonula occludens*) are normally the most apical of the junctions and form continuous ribbon-like structures around the cell. One or more fusion sites may be visible between the adjacent membranes. The function of these fusion sites is to prevent the flow of material (including ions and water) between the cells. The more sites which occur, the greater the impermeability of the epithelia. For example, proximal tubule cells contain less fusion sites than the lining epithelial cells of the bladder and consequently are more permeable to small molecules. Like tight junctions, belt desmosomes (*zonulae adherentes*; singular, *zonula adherens*) also encircle the cell. These intercellular junctions are characterized by dense plaques on the inner surface of both plasma membranes into which merge actin-containing microfilaments arising from the cytoskeleton.

Gap junctions occur randomly along the lateral surface of the cells and are characterized by the close apposition of adjacent cell membranes. Hydrophilic channels connect the cells and allow the transfer of substances of $M_w \leq 1500$. Thus molecules, including second messengers (e.g. cyclic AMP and Ca^{2+} ions), can spread amongst the cells of a tissue and coordinate their behaviour.

Spot desmosomes (*maculae adherentes*; singular, *macula adherens*) resemble belt desmosomes in that dense plaques are present on the inner surface of both cell membranes, but, like gap junctions, they appear as discrete zones on the lateral cell membrane. Keratin filaments from the cytoskeleton attach to the proteinous plaques and thus greatly enhance the strength of the epithelial tissue. Most epithelial cells are bound to the basement membrane by hemidesmosomes which resemble half desmosomes, though these structures are absent from certain epithelia, e.g. mesothelia.

1.7 Basement membrane

Where epithelial tissues border connective tissues (e.g. the epidermal–dermal junction) or endothelium (e.g. in the kidney glomerulus and lung alveoli), a thin layer of specialized extracellular matrix appears called a basement membrane. Electron micrographs reveal that this structure contains an electron-dense layer (lamina densa) normally less than 100 nm thick, which runs parallel to the basolateral membranes of the epithelial cells. The lamina densa is straddled by electron-lucent layers referred to as the lamina lucida (on the epithelial and endothelial sides) and the lamina fibroreticularis (on the connective tissue side). The basement membrane fulfils three essential roles:

(a) regulation of the passage of macromolecules to and from the epithelial tissues;
(b) provision of a flexible matrix for epithelial cells to attach to; and
(c) endowment of polarity on the epithelial cells through interactions with basal receptors.

The lamina densa comprises a meshwork of type IV collagen which, unlike type I and type III collagen, is not characteristic of loose connective tissue. This component along with others, such as the glycoprotein laminin, and proteoglycans, are products of the cells which the basement membrane underlies (i.e. epithelial and/or endothelial cells). In contrast, the reticular fibres of the lamina fibroreticularis contain type III collagen, a product of the fibroblasts of the connective tissue. Laminin is important for adhering epithelial cells to the basement membrane as it has affinity for both receptors on the basal surface of the cells and to type IV collagen and proteoglycans. Likewise, fibronectin, another high molecular weight constituent of basement membranes, couples cell membrane receptors to various basement membrane components. Fibronectin also binds to type III collagen, thus securing the basement membrane to the underlying reticular fibres.

2. Epithelia: methods of culture

2.1 Introduction

The cultivation of many epithelial cell types *in vitro* truly came of age in 1975 when James Rheinwald and Howard Green introduced the use of feeder layers of irradiated mouse 3T3 mesenchymal cells for culturing keratinocytes (see Chapter 9, *Protocol 4C*; refs 5, 6). These mouse fibroblasts assist the growth of epithelial cells while suppressing the growth of other cell types, most notably other fibroblasts. Rheinwald and Green also recognized that keratinocytes require specific growth factors not present in serum, e.g. epidermal growth factor (EGF), and certain growth conditions, e.g. small colony size, in order to maintain them in a proliferating state (7, 8). These conditions, for the first time, allowed many epithelial cell types to be expanded in culture. It should be noted, however, that not all epithelial cells are conducive to this system, e.g. hepatocytes and enterocytes (9).

The ability to culture epithelial cells has been invaluable for studying the cellular mechanisms of this cell type. Indeed, culture techniques have provided many advantages over whole animal experimentation including superior control of the physicochemical environment and physiological conditions (especially where serum-free medium can be used; ref. 10), the opportunity to study human cells, and unprecedented accessibility for biochemical monitoring and observation. Furthermore, they provide a more ethical, and sometimes a cheaper alternative, to *in vivo* studies. However, when cells are cultured and propagated as described above (i.e. in conditions designed to encourage cell proliferation, not cell differentiation), the resultant cell phenotype differs dramatically with increasing passages from the characteristics predominant in the original tissue. In consequence, these cells are significantly different in morphology, biochemistry, and function from their counterparts *in vivo*, and thus are far from ideal for studies of specialized cell behaviour. To circumnavigate this problem, two *in vitro* alternatives exist; organ culture or histotypic/organotypic cell culture (see *Table 3* for definition of terminology).

2.2 Organ culture

Where maintenance of differentiated characteristics is required, organ culture has proved particularly useful (11). In these cultures, cells are retained within a three-dimensional structure displaying correct spatial organization and intact cell–matrix and homotypic and heterotypic cell–cell interactions.

Table 3. Definition of culture techniques

Organ culture
A technique whereby small, undisaggregated tissues are cultured at an air–liquid interface so that they retain a 3D structure and some or all of the histological and functional features of the tissue *in vivo*.

Cell culture
A technique whereby isolated cells taken from a tissue, primary explant, or from a cell line or cell strain, by enzymatic, chemical, or mechanical disaggregation, are cultivated as suspensions or as adherent layers.

Histotypic cell culture
A technique whereby cells are isolated as described above and are reassociated in a manner so that they resemble a tissue of the organ under study.

Organotypic cell culture
A technique whereby cells of different lineages are isolated as described above and are recombined so that they recreate one or more tissues of the organ under study.

Epithelial tissues invariably follow a pattern of differentiation *in vitro* analogous to that *in vivo* and can, for example in glandular or bronchial tissues, develop new structures at the periphery of the culture. Unfortunately however, organ cultures can bring with them a number of problems which can severely limit their usefulness:

(a) The presence of numerous cell types, in addition to those of interest, may make the system unsuitable for cellular and molecular studies.

(b) Organ cultures cannot be propagated and so there is a continual requirement for new tissue. This can be a major problem where human tissue is being used.

(c) Reproducibility between cultures can be poor, especially where more than one donor has been used.

2.3 Histotypic and organotypic cell culture

Epithelial tissues are an equilibrium between undifferentiated cells, which normally have a high capacity to proliferate (i.e. stem cells and committed precursor cells) and differentiated cells, which have a low capacity to proliferate. Furthermore, this equilibrium is rigorously controlled so that tissue architecture and function are preserved. However, the abnormal environment in which isolated cells have been traditionally cultured (i.e. at a low cell density and in the presence of serum and/or growth factors) grossly alters this equilibrium by encouraging the proliferation of unspecialized cells.

Specialized functions are usually expressed most strongly in primary cultures, especially where the cultures have become confluent; such conditions reduce proliferation and may encourage differentiation through increased formation of tight and gap junctions, and changes in cell shape. The progressive loss of differentiated functions with passage normally results from a combination of:

(a) undifferentiated cells (of the same or another lineage) overgrowing the less prolific mature cells; and

(b) the absence of appropriate inducers of differentiation.

Selective culture techniques (i.e. the use of 3T3 feeder layers and serum-free media) have largely removed the first of these problems. The second problem, of inducing cells to differentiate, is being addressed through a better understanding of the factors which govern cell differentiation, namely the supporting matrix, the media components, and cell–cell interactions (12). Histotypic and organotypic cultures attempt to optimize these elements to recreate an environment for the cells which more closely resembles that experienced *in vivo*, and is therefore conducive to the expression of the differentiated phenotype.

2.3.1 The matrix

As discussed in Section 1.7, epithelial cells *in vivo* rest on a basement membrane which regulates the differentiation of these cells. The principal features of this membrane are that it is permeable and that it is the route via which epithelial cells receive their nutrients. These attributes can easily be recreated *in vitro* by culturing epithelial cells on a porous substrate through which media can permeate, e.g. polycarbonate or nitrocellulose filters (see Chapters 3 and 5–8). This simple modification of conventional (solid surface) cell culture dramatically alters the morphology of most epithelial cells: for example, the immortal cell lines, MDCK (see Chapter 3) and Caco-2 (see Chapter 6), appear squamous when grown on solid plastic and display poor cell polarity. However, when these cells are grown on inert filters, their morphology dramatically alters to become cuboidal and columnar, respectively, and they both display good cell polarity which is essential for developing epithelial differentiation. The saturation density of the cells is 3–4 times greater on porous substrates than on solid plastic or glass.

However, the basement membrane is more than just a porous substrate through which the cells receive their nourishment. It is also important in providing the cells with a flexible support and for interacting with basolateral receptors to regulate cell differentiation. Primary epithelial cell cultures generally require these cell–matrix interactions if they are to maintain their differentiated properties. This feature can be modelled, in part, by:

(a) coating inert filters with one or more extracellular matrix components, e.g.
   i. collagen; type I, type I/III, or type IV
   ii. fibronectin (see Chapter 5)
   iii. laminin
   iv. natural extracellular matrix, e.g. Matrigel (see Chapters 7 and 8);

(b) using artificial or natural substrates comprising one or more extracellular matrix components, e.g.
   i. collagen gels (see Chapter 9)
   ii collagen–glycosaminoglycan membranes (13)
   iii. tissue denuded of epithelia (see Chapter 9)
   iv. *in vivo* transplantation sites (14, 15).

2.3.2 Media components and cell–cell interactions

The composition of the media is important for regulating cell proliferation and differentiation, and thus for forming and maintaining a tissue-like architecture *in vitro*. Numerous soluble factors, including endocrine hormones (e.g. hydrocortisone), vitamins (e.g. retinoic acid; refs 16–18), inorganic ions (e.g. Ca^{2+} ions; ref. 19), and non-physiological substances (e.g. phorbol esters;

ref. 20) can control the differentiation of various epithelial cells. Furthermore, cell behaviour can be markedly influenced by extracellular matrix components and paracrine factors (e.g. TGF-β; ref. 21) released from a second neighbouring cell type (15). These factors can be used to regulate cell differentiation in culture by:

(a) adding them directly to the cell culture medium;

(b) using media conditioned by the factor-producing cell type (e.g. astrocyte-conditioned medium; see Appendix 2); or

(c) co-culturing the factor-producing cell type with the epithelial cells (e.g. fibroblasts; see Chapters 7 and 9).

2.3.3 Discussion

Histotypic and organotypic cell culture systems attempt to amalgamate the benefits offered separately by cell and organ culture methods. In many instances these novel cell cultures, like organ cultures, closely mimic epithelia *in vivo*; cells display a tissue-like organization, develop distinct apical and basolateral surfaces, and retain many differentiated functions, e.g. vectorial transport (see Chapters 3 and 6) and secretion (see Chapters 7 and 8), cytokeratin expression (see Chapter 2) and xenobiotic metabolism (see Chapter 4), which are missing or under-expressed in conventional cell cultures. These and other characteristics can be assessed to compare the similarity of these models to their *in vivo* counterparts (see *Table 4*; refs 22–28). However, these systems are distinct from organ cultures in that they contain a limited number of cell types (which may be advantageous for cellular and molecular studies), and the cells can usually be propagated, thus reducing or negating the need for donor tissue. This latter point is important because it increases the opportunity to use human cells and improves the reproducibility of intra- and inter-laboratory cultures. Furthermore, the opportunity to manipulate the cellular environment and to use normal and/or diseased cells (29, 30) affords the chance to study cell–matrix and cell–cell interactions, and the effect of disease on the processes of cell proliferation and differentiation.

The use of histotypic and organotypic cell culture systems is expanding rapidly but is still in its infancy. The models detailed in the following chapters consider a variety of different epithelia and the diverse studies to which they can be applied. Furthermore, these examples aptly demonstrate the basic principles behind the development of these models which have been presented in this chapter. Researchers should be aware, however, that these systems can be difficult to set up, especially for those with little or no previous experience of cell culture methods. For those bold enough to develop their own models, a number of fundamental questions should be addressed:

(a) Do you require a system where the cells display specialized functions and if so, would other techniques be more appropriate, e.g. organ culture?

Table 4. Comparison of epidermis *in vivo* and epidermis reconstituted on de-epidermized dermis *in vitro*[a]

Morphology		***In vivo***	***In vitro***
Strata: stratum basale		✓	✓
stratum spinosum		✓	✓
stratum granulosum		✓	✓
stratum corneum		✓	✓
Basement membrane		✓	✓ (incomplete)
Hemidesmosomes		✓	✓
Desmosomes		✓	✓
Keratohyaline granules		✓	✓
Membrane coating granules		✓	✓
Cornified cell envelope		✓	✓
Function			
Physical barrier to xenobiotics		✓	✓ (impaired)
Metabolism of xenobiotics		✓	✓[b]
Protein markers of differentiation			
S. basale:	keratins	K5/K14	K5/K14
	BP antigen	✓	✓
S. spinosum:	keratins	K1/K10	K1/K10, K6/K16[c]
	filaggrin	x	✓[c]
	involucrin	x	✓[c]
	transglutaminase	x	✓[c]
S. granulosum:	filaggrin	✓	✓
	involucrin	✓	✓
	transglutaminase	✓	✓
Lipid markers of differentiation			
Ceramides		✓	✓
Essential fatty acids		✓	✓ (low)
Lanosterol		✓	✓
Sphingolipids		✓	✓
Triglycerides		✓	✓ (high)

[a] Summary of references 22–28.
[b] Keratinocytes were grown on a collagen-fibroblast lattice (28).
[c] Characteristic of hyperproliferative (regenerative) skin.

(b) Should you use human or animal cells?

(c) Should you use primary cultures or cell lines? Some immortal tumourigenic (e.g. Caco-2; see Chapter 6) and non-tumourigenic cell lines (e.g. MDCK; see Chapter 3) retain many differentiated characteristics and are suitable for histotypic cell culture. The use of such cell lines avoids many of the problems entailed with primary cultures.

(d) Should you use embryonic cells? Although embryonic cells usually have a longer 'culture life' than adult cells, presumably because of a higher percentage of stem cells in the population, they don't necessarily develop in culture into fully mature cells.

(e) Do the cells require a cell matrix?

(f) Is it worth co-culturing the cells with a second cell type?

References

1. Junqueira, L. C., Carneiro, J., and Kelley, R. O. (ed.) (1992). *Basic histology* (7th edn). Appleton & Lange, (Prentice Hall Int.), London, UK.
2. Cormack, D. O. (1987). *Ham's histology* (9th edn). J. B. Lippincott Co., Philadelphia, PA.
3. Moll, R., Franke, W. W., Schiller, D. L., Geiger, B., and Krepler, R. (1982). *Cell*, **31**, 11.
4. Cooper, D., Schermer, A., and Sun, T-T. (1985). *Lab. Invest.*, **52**, 243.
5. Rheinwald, J. G. and Green, H. (1975). *Cell*, **6**, 317.
6. Rheinwald, J. G. and Green, H. (1975). *Cell*, **6**, 331.
7. Rheinwald, J. G. and Green, H. (1977). *Nature*, **265**, 421.
8. Rheinwald, J. G. (1979). *Int. Rev. Cytol.*, **10** (Suppl.), 25.
9. Rheinwald, J. G. (1989). In *Cell growth and division: a practical approach* (ed. R. Baserga), pp. 81–94. IRL Press, Oxford.
10. Maurer, H. R. (1992). In *Animal cell culture: a practical approach* (2nd edn) (ed. R. I. Freshney), pp. 15–46. IRL Press, Oxford.
11. Lasnitski, I. (1992). In *Animal cell culture: a practical approach* (2nd edn) (ed. R. I. Freshney), pp. 213–61. IRL Press, Oxford.
12. Freshney, R. I. (1992). In *Culture of epithelial cells* (ed. R. I. Freshney), pp. 1–23. Wiley–Liss, Inc., NY.
13. Shahabeddin, L., Berthod, F., Damour, O., and Collombel, C. (1990). *Skin Pharmacol.*, **3**, 107.
14. De Luca, M., Albanese, E., Megna, M., Cancedda, R., Mangiante, P. E., Cadoni, A., and Franzi, A. T. (1990). *Transplantation*, **50**, 454.
15. Fusenig, N. E. (1992). In *Culture of epithelial cells* (ed. R. I. Freshney), pp. 25–57. Wiley–Liss, Inc., NY.
16. Wu, R. and Wu, M. M. (1986). *J. Cell Physiol.*, **127**, 73.
17. Asselineau, D., Bernard, B. A., Bailly, C., and Darmon, M. (1989). *Dev. Biol.*, **133**, 322.
18. Regnier, M. and Darmon, M. (1989). *In Vitro Cell Dev. Biol.*, **25**, 1000.
19. Boyce, S. T. and Ham, R. G. (1983). *J. Invest. Dermatol.*, **81** (suppl.), 33s.
20. Willey, J. C., Saladino, A. J., Ozanne, C., Lechner, J. F., and Harris, C. C. (1984). *Carcinogenesis*, **5**, 209.
21. Coffey Jr., R. J., Sipes, N. J., Bascom, C. C., Graves-Deal, R., Pennington, C. Y., Weissmann, B. E., and Moses, H. L. (1988). *Cancer Res.*, **48**, 1596.
22. Ponec, M., Weerheim, A., Kempenaar, J., Mommaas, A. M., and Nugteren, D. H. (1988). *J. Lipid Res.*, **29**, 949.
23. Ponec, M. (1991). *Toxicol. In Vitro*, **5**, 597.
24. Asselineau, D., Bernard, B. A., Bailly, C., and Darmon, M. (1989). *Dev. Biol.*, **133**, 322.

25. Ponec, M., Wauben-Penris, P. J. J., Burger, A., Kempenaar, J., and Bodde, H. E. (1990). *Skin Pharmacol.*, **3**, 126.
26. Mak, V. H. W., Cumpstone, M. B., Kennedy, A. H., Harmon, C. S., Guy, R. H., and Potts, R. O. (1991). *J. Invest. Dermatol.*, **96**, 323.
27. Regnier, M., Asselineau, D., and Lenoir, M. C. (1990). *Skin Pharmacol.*, **3**, 70.
28. Pham, M-A., Magdalou, J., Siest, G., Lenoir, M. C., Bernard, B. A., Jamoulle, J. C., and Shroot, B. (1990). *J. Invest. Dermatol.*, **94**, 749.
29. Regnier, M., Desbas, C., Bailly, C., and Darmon, M. (1988). *In Vitro Cell Dev. Biol.*, **24**, 625.
30. De Dobbeleer, G., de Graef, Ch., M'Poudi, E., Gourdain, J. M., and Heenen, M. (1989). *J. Am. Acad. Dermatol.*, **21**, 961.

2

Identification and localization of cytokeratins

IAN C. MACKENZIE and ZHIRONG GAO

1. Introduction

1.1 Organotypic cultures and epithelial differentiation patterns

The patterns of polarization, stratification, and maturation that characterize epithelial differentiation *in vivo* are usually lost when epithelial cells are grown on plastic substrates using standard cell culture methods. However, epithelia cultured on organic matrices containing mesenchymal cells often redevelop regionally specific patterns of growth and structure, and such culture techniques thus provide unique opportunities for the *in vitro* investigation of many aspects of epithelial differentiation and function. The demonstrable effects of mesenchyme on epithelial differentiation in organotypic cultures illustrate the importance of tissue interactions in the maintenance of normal epithelial structure. Organotypic cultures provide valuable systems for the study of the actions and molecular nature of these mesenchymal influences.

Although histological or ultrastructural studies provide useful information about epithelial growth and differentiation, many types of studies require more detailed analysis of patterns of epithelial macromolecular synthesis. These are sometimes necessary to characterize fully or to standardize an organotypic system being used for functional studies, but such analysis also forms a useful way of detecting various experimentally induced changes. Different types of epithelia, for example simple and stratified epithelia, markedly differ in their patterns of macromolecular expression in a way that reflects regionally differing functions such as secretion, barrier formation, or distensibility. Certain molecules are expressed only in certain epithelia (e.g. casein by mammary epithelium, plakin by bladder epithelium), but there are families of structurally related molecules, such as the cytokeratins, that are ubiquitously expressed. Carbohydrate-bearing molecules of the blood group antigen series (1), cell adhesion molecules such as integrins (2), and various molecules associated with basal lamina formation (3) represent other families of molecules whose patterns of cellular expression in epithelia vary both with epithelial origin and with the stage of maturation. However, the cytokeratins have been extensively studied and found to be particularly useful markers of normal and experimentally altered patterns of epithelial differentiation (4–6).

1.2 The cytokeratins

Molecules forming intermediate filaments are grouped into several families that usually, but with several exceptions, are expressed in a cell lineage specific fashion: for example, vimentin is expressed by mesenchymal cells, such as fibroblasts, and cytokeratins by epithelial cells. First comprehensively catalogued by Moll *et al.* (4), the cytokeratins are classified into basic and acidic subfamilies and numbered as cytokeratins #1–9 (basic) and #10–19 (acidic) according to their patterns of mobility during two-dimensional gel elec-

Table 1. Basic patterns of cytokeratin distribution and a list of antibodies with well-defined specificities[a]

Cytokeratin	Common sites of expression	Antibody	Supplier	Reference
5	Basal cells in all stratifying epithelia.	Basal 2	ICM	5
14		LL002	Biogenesis	5
1	Suprabasal cells of epithelia cornifying	—	—	
10		RKSE60	ICN Biomedicals	17
4	Suprabasal cells of epithelia non-cornifying	6B10	ICN Biomedicals	18
13		1C7	Biogenesis	18
3	Corneal epithelia	AE5	Biogenesis	19
12				
6	Hyperplastic epidermis, epithelia some mucosal	—	—	
16		LL025	Dr E. B. Lane	
7	Simple epithelia	RCK 105	ICN Biomedicals	20
8		LE 41	Amersham Int.	21
17		E3	Thamer	22
18		LE 61	Dr E. B. Lane	20
19	Simple epithelia, basal in some mucosal epithelia	LP2K	Amersham Int.	23
Pan-keratins	Epithelia in general	LP34	DAKO	

[a] Cytokeratins are listed (as the typically co-expressed acidic/basic pairs) by distribution in tissues and numbered according to Moll (4). Each antibody listed has a defined specificity for a human cytokeratin as described in the reference listed. Each of these antibodies was selected from a wide range tested for (a) cytokeratin specificity and low cross-reactivity with any non-cytokeratin components, (b) their apparent recognition of murine cytokeratins corresponding to those recognized in the human catalogue.

trophoresis. A cytokeratin expressed in simple epithelia (#20), and several more cytokeratins that are expressed in structures such as hair or nail (7), have subsequently been added to this catalogue. The formation of cytokeratin filaments requires heteropolymerization of cytokeratins from each of the acidic and basic subfamilies: all epithelial cells therefore express at least two cytokeratins. However, the cytokeratins are unusual among intermediate filament molecules in their degree of molecular diversity and, for reasons which are not fully elucidated, most epithelia show more complex patterns of expression. A highly simplified summary of the basic patterns of epithelial expression is shown in *Table 1*.

A great deal of information is now available regarding the molecular structure of cytokeratins and the organization and control of expression of cytokeratin genes. For general information about cytokeratins, and their normal and pathological expression in particular tissues, the reader is advised to refer initially to recent review articles (e.g. ref. 6). The aims of this chapter are restricted to describing general methods for examining the epithelial expression of cytokeratins using either immunohistochemistry or electrophoresis. Each of these methods has certain advantages and disadvantages, but, individually or in combination, they can reliably provide basic information about the patterns of tissue expression of cytokeratins. Other molecular biological techniques, including *in situ* hybridization, can provide further information about cytokeratin expression at the mRNA level but will not be discussed here. The methods we have described below are those routinely used in our laboratory. The sources of materials listed are not comprehensive: they simply represent those we use and with which we have achieved acceptable results.

2. Immunohistochemical techniques for the demonstration of cytokeratins

2.1 General considerations

Cytokeratins are relatively abundant, stable, and insoluble cellular components and are, therefore, quite readily demonstrated by immunohistochemistry. Antibodies with defined specificities for many of the individual human cytokeratins are now available and permit relatively rapid and reliable assessment of cytokeratin expression in human tissues. The primary advantage of immunohistochemical techniques is that they permit accurate localization of cytokeratin expression in individual cells, or in groups or strata of cells, and can thus provide information about the homogeneity of cell populations, and about sequential patterns of cell maturation, not readily obtainable by other methods. The basic procedures for demonstrating cytokeratins by immunohistochemistry consist of:

(a) appropriate preparation of tissues;

(b) selection of appropriate antibodies for the demonstration of cytokeratins of interest;

(c) establishment of conditions in which primary antibodies bind specifically to their target epitopes;

(d) demonstration of the binding of primary antibodies, usually by secondary antibodies conjugated to agents capable of generating a visible signal;

(e) establishment of suitable controls for the specificity of the observed signals.

As a general rule, the immunocytochemical demonstration of cytokeratins, in either organotypic cultures or normal control tissues, is most consistently achieved using unfixed frozen tissue sections. Structural detail may be better preserved using fixed, wax-embedded tissues, but almost always at the expense of loss of effective primary antibody binding. To characterize cell populations being used to establish organotypic cultures, cells growing as monolayers on tissue culture plastic can be stained directly *in situ*. Alternatively, cell populations can be stained as cell smears after removal from their substrate.

2.2 Preparation of samples for immunohistochemistry

2.2.1 Preparation and sectioning of frozen tissues

Specimens need to be carefully oriented for sectioning, usually perpendicular to the epithelial surface. The method (8) outlined below allows precise orientation as well as convenient handling, labelling, and storage of specimens.

(a) Label a cork disc with indelible ink and make a shallow vertical incision across its diameter. Place the specimen on a plastic coverslip inserted into this incision, cover with OCT compound (Baxter) and appropriately position, flatten, and orientate the specimen by manipulation with fine forceps.

(b) Freeze the specimen by immersion in isopentane cooled in liquid nitrogen. Use isopentane to avoid boiling of the nitrogen which slows the rate of freezing and can disturb the specimen.

(c) Remove the coverslip (by warming slightly with a fingertip) to leave the specimen oriented perpendicular to the disc and ready for sectioning.

(d) Store the mounted specimens indefinitely in liquid nitrogen or for shorter periods in a –80 °C freezer (seal in a plastic bag with an ice cube to prevent desiccation).

(e) For sectioning, use OCT compound to mount the cork disc on a cryostat chuck and section in a cryostat at about 6 μm. Collect sections on cleaned microscope slides and briefly air-dry, preferably in a laminar-flow hood. Either stain immediately or store slides bearing sections in slide boxes in a –80 °C freezer. If the specimen is small enough, collect sections on 10-

well slides (Cell Line Associates) to allow a range of control and other staining reactions on the same slide. Alternatively, groups of two or three sections can be collected on ordinary slides and isolated for staining by circling with a 'PAP' pen (Binding Site).

2.2.2 Preparation of cells growing on plastic

(a) Briefly wash the dish or flask in three changes of PBS: 8 g NaCl, 0.2 g KCl, and 1.44 g Na_2PO_4 dissolved in 800 ml of double-distilled water. Adjust the pH to 7.2, and make up to 1 litre using double-distilled water.

(b) Shake off the PBS and air-dry.

(c) Cut out appropriately sized pieces (i.e. 2 × 5 cm) of the plastic bearing the cells by using a hot instrument (e.g. electric soldering iron or spatula heated in a flame).

(d) Permeabilize the cells for antibody access by immersion in ice-cold acetone:methanol (1:1) for 2 minutes. This is sometimes referred to as 'fixation', but this process neither cross-links nor coagulates proteins; its main function is to remove lipid and open up the cell membrane. Test that acetone does not attack the particular type of tissue culture plastic being used. If it does, either treat with methanol alone or stain 'unfixed' but freeze and thaw the specimen to permeabilize the cells.

(e) Mark a grid on the dish with a PAP pen to isolate several areas of the cell population for staining with different antibodies and controls.

2.2.3 Preparation of cell smears

(a) Take an aliquot of cells when they are released from the culture vessel by trypsinization.

(b) Centrifuge to pellet the cells, remove the supernatant, and resuspend the cells in a single drop of 20% calf serum in PBS.

(c) Using as little fluid as possible, smear the cells on standard or 10-well slides and allow them to air-dry rapidly (if a lot of fluid is used, increasing osmolarity and crystallization of salts as the fluid evaporates can damage the cells and reduce adhesion).

(d) "Fix" cell smears in cold (4 °C) acetone:methanol (1:1).

2.3 The use of processed tissues for immunohistochemistry

Antibody recognition of cytokeratins in processed tissues is almost always greatly reduced by the formation of cross-links and altered protein conformation during fixation, dehydration, and heating. Tissue processing may be unavoidable for some studies, for example when it is necessary to use a single specimen to localize both soluble or unstable tissue components and cytokeratins. The choice of fixation will usually be determined by the properties of non-cytokeratin components, but fixation in freshly made buffered formaldehyde is a good general purpose method (4 g paraformaldehyde plus 3 drops 1 M NaOH plus 100 ml PBS; heat on a hot plate/stirrer until clear. Cool, filter and adjust the pH to 7.2). Times for fixation will depend upon the size of the specimens and can range from a few minutes at 4 °C for very small pieces to overnight for larger pieces. After fixation, wash the specimens in several changes of PBS and then freeze for sectioning (as in Section 2.2.1) or embed in paraffin wax by a 'cold' processing technique, e.g. Sainte-Marie (9), which avoids undue heating of the specimen during embedding. Care should also be taken not to overheat sections when mounting them on to microscope slides.

Antigens blocked during processing can often be partially restored by dewaxing and rehydrating the sections and then treating them with proteolytic enzymes, (such as 0.1% (w/v) trypsin for 30 minutes at 37 °C), or with commercially available methods (e.g. Antigen Retrieval System, Biogenex). The latter are more consistent and usually provide acceptable results for generalized epithelial staining with polyclonal or pan-specific antibodies. However, we have found none of these methods to be fully effective; for example, epitopes that are recognized by some monoclonal antibodies may be partially restored, but others not. Therefore, be wary of false-negative results, and test for the recovery of the epitopes recognized by each type of mAb by processing and staining control tissues known to contain cytokeratins of interest.

2.4 Selection of primary antibodies

Structural variation between individual cytokeratins provides regions of unique structural sequence that are available for specific recognition by antibodies. Monoclonal antibodies with defined specificities for most of the individual human cytokeratins are now available either commercially or from colleagues (see *Table 1*). Other mAbs, such as LP34 (10), recognize epitopes that are present on several cytokeratins and thus act as 'pan-specific' markers for all types of cytokeratin-expressing epithelium. Yet other mAbs recognize epitopes shared only by subgroups of cytokeratins, e.g. by members of the acidic or basic subfamilies (11).

Polyclonal and monoclonal anti-cytokeratin antibodies tend to differ somewhat in their specificities and uses. Monoclonal antibodies, recognizing single epitopes on the molecules against which they are raised, are very specific and, as they can be produced in large quantities, their specificities can be well defined. If mAbs with appropriate specificities are available, they are effective for the identification of individual cytokeratins. However, mAbs are sensitive to the effects of factors such as cross-linking fixatives which alter molecular conformation. Polyclonal antibodies typically contain a spectrum of antibodies recognizing numerous epitopes on the molecules against which they are raised. They may be less specific than mAbs but of more value in detecting cytokeratins in processed tissues or in a denatured form, as in Western blots. Polyclonal antibodies also usually contain irrelevant antibodies that were present in the blood of the immunized animal and which may cross-react with tissue components other than cytokeratins: greater care may therefore be necessary in choosing controls for their specificity. However, polyclonal antibodies are much simpler than mAbs to raise against defined epitopes such as synthetic peptides, and problems concerning the specificity of polyclonal antibodies can often be dealt with by immunoabsorption or other methods.

At present, there are few antibodies with well-defined specificities for individual cytokeratins of non-human species. The regional patterns of expression of particular cytokeratins tend to be similar for different species, and there is some homology between individual cytokeratins of various mammals. The mAbs that recognize an individual human cytokeratin may, therefore, also recognize a corresponding cytokeratin in other species. The mAbs listed in *Table 1* have been selected because they not only specifically recognize individual human cytokeratins but also appear to recognize corresponding cytokeratins in murine tissues. However, some cytokeratins in each species may diverge from the human catalogue: for example, the mouse expresses several cytokeratins not directly corresponding to those found in humans (12). When specific mAbs for non-human cytokeratins cannot be obtained, cytokeratin expression can be studied by gel electrophoresis. However, it may be possible to raise antibodies for particular experimental purposes and, for a more extended discussion of the properties, preparation, and purification of specific monoclonal and polyclonal antibodies, the reader is referred to texts such as Harlow and Lane (13).

2.5 Detection of antibody binding

Anti-cytokeratin antibodies can be directly conjugated or 'tagged' with accessory molecules that enable binding to be visually displayed, but this is seldom a practical choice. An indirect method, in which binding of a primary anti-cytokeratin antibody is detected by a secondary antibody conjugated with a fluroescent or other label, is almost always preferable. A wide range of, usually polyclonal, secondary antibodies directed against immunoglobulins of various species and conjugated to various types of label is commercially available (e.g. DAKO). The use of the indirect technique has the advantages of amplifying the signal produced by the primary antibody and permitting the use of one conjugated secondary antibody to detect all primary antibodies having the same species origin.

Fluorescein isothiocyanate (FITC) is a commonly used fluorescent label which produces a strong, clean, well-localized signal. FITC avoids the need for the additional steps necessary to display enzyme labels, but, like other fluorescent labels, has the disadvantage of requiring a microscope suitably equipped with a UV source and appropriate barrier filters. FITC also tends to fade when excited by the UV beam, but fading can be largely controlled by adding agents such as phenylenediamine to the mounting medium (14). FITC requires the use of aqueous mountants: these do not provide permanent specimens but can be stored at –80 °C without loss of quality for quite a long time. To examine FITC-conjugated antibodies, simply cover the stained sample in aqueous mountant (100 mg *p*-phenylenediamine in 10 ml PBS added to 90 ml glycerol; pH adjusted with NaOH to pH 8.0).

Secondary antibodies conjugated to a variety of enzymes such as horseradish peroxidase, alkaline phosphatase, or β-galactosidase are also widely available (e.g. DAKO), and their binding is displayed by a histochemical technique appropriate for the particular enzyme. Enzyme-linked secondary antibodies produce a strong, permanent signal, but their use may be associated with higher non-specific background staining. It is also necessary to block any endogenous enzyme that may be present in the tissue being examined.

2.6 Controls for specificity of staining reactions

The signal observed as the result of an immunohistochemical staining reaction may not indicate specific recognition of the molecule against which the primary antibody is directed. Reasons for this include:

(a) cross-reactivity of the primary antibody with epitopes present on other tissue components;

(b) non-specific binding of primary or secondary antibodies to tissue components;

(c) the detection of endogenous enzyme by processing procedures for enzyme-labelled antibodies.

Most mAbs have well-defined specificities and have been tested for cross-reactivity with non-cytokeratin tissue components. Polyclonal antibodies are often less well characterized and are more likely to show cross-reactivity. The simplest way to check each antibody is by staining various control tissues for comparison of the known and the observed distributions of cytokeratins.

Non-specific binding of antibodies can be tested by performing the full staining procedure and:

(a) using an irrelevant mAb of the same isotype as the specific mAb being used, or pre-immune serum from the animal in which a polyclonal antibody was raised, or

(b) omitting the primary antibody.

Any signal detected is due to non-specific binding. If enzyme-linked secondary antibodies are used, the enzyme histochemistry procedure should be performed on sections that are not exposed to antibody but are treated to display enzyme activity, both blocked and unblocked for the activity of endogenous enzyme. Non-specific binding of antibodies can be reduced primarily in two ways:

(a) A high concentration of antibodies produces a background of non-specific binding, therefore this can be reduced by increasing the dilution of the antibodies; use a chequer-board of primary and secondary antibodies to determine the highest dilutions that produce a satisfactory signal. It is also more cost-effective to use high dilutions of commercial antibodies.

(b) Charged sites on tissue components may bind antibodies non-specifically. Pretreatment of sections with proteins such as bovine serum albumin or calf serum, and maintenance of high levels of such protein in the staining and washing solutions, can block charged sites and reduce the background signal.

The following staining procedure (*Protocol 1*) is suitable for frozen sections and cell smears (see section 2.2) and also for processed tissues after treatment for antigen re-exposure as described in Section 2.3. An example of the use of immunofluorescent staining of organotypic cell cultures is displayed in *Figure 1*.

Protocol 1. General staining procedures

Equipment and reagents

- Primary antibodies directed against cytokeratin(s) of interest (see Section 2.4)
- Secondary antibodies conjugated to chosen detection system (see Section 2.5)
- PBS with 0.5% bovine serum albumin (pH 7.2)
- Coverslips and mounting media
- Humid incubation chamber; place damp tissues on the base of a lidded box
- Enzyme substrates as necessary (see Section 2.5), e.g. AP substrate working solution: 2 mg naphthol AS-MX phosphate (Sigma cat. no. N-5000), 10 ml of 0.1 M Tris buffer (pH 7.8), and 2.4 mg levamisole (levamisole inhibits most tissue phosphatases but not the intestinal phosphatase used to label secondary antibodies). Add 10 mg Fast Red TR salt (Sigma, cat. no. F–1500) just before use

Method

1. Bring stored frozen slides or dishes to room temperature and dry briefly in a hood.
2. Cover each sample with the primary antibody and incubate for at least 2 h at room temperature or, preferably, overnight at 4°C.[a] A volume of about 30 μl of primary antibody is suitable for most specimens. Adjust the volume for large sections.
3. Shake off the primary antibody, wash the specimen three times over 15 min in PBS, and shake off excess fluid.
4. Apply secondary antibody specific for the immunoglobulin of the species in which the primary antibody was raised (e.g. mouse, rat, goat, etc.) and incubate at room temperature for 1 h.
5. Wash in three changes of PBS over 15 min and process according to the conjugate of the secondary antibody (see Section 2.5).[b,c]

[a] It is important that the specimens are not allowed to 'dry out' at any stage after the primary antibody is added.

[b] To display alkaline phosphatase conjugated antibodies: rinse the tissue in Tris buffer for

Protocol 1. *Continued*

10 min, then incubate in alkaline phosphatase substrate working solution for 10–20 min until the colour develops. Rinse in buffer, dehydrate in graded alcohols, and cover with glycerin jelly and a coverslip.

[c] To display peroxidase conjugated antibodies: use an LSAB substrate kit (DAKO) according to the manufacturer's instructions (includes a step for blocking endogenous tissue peroxidases with peroxide). Dehydrate and mount in Permount.

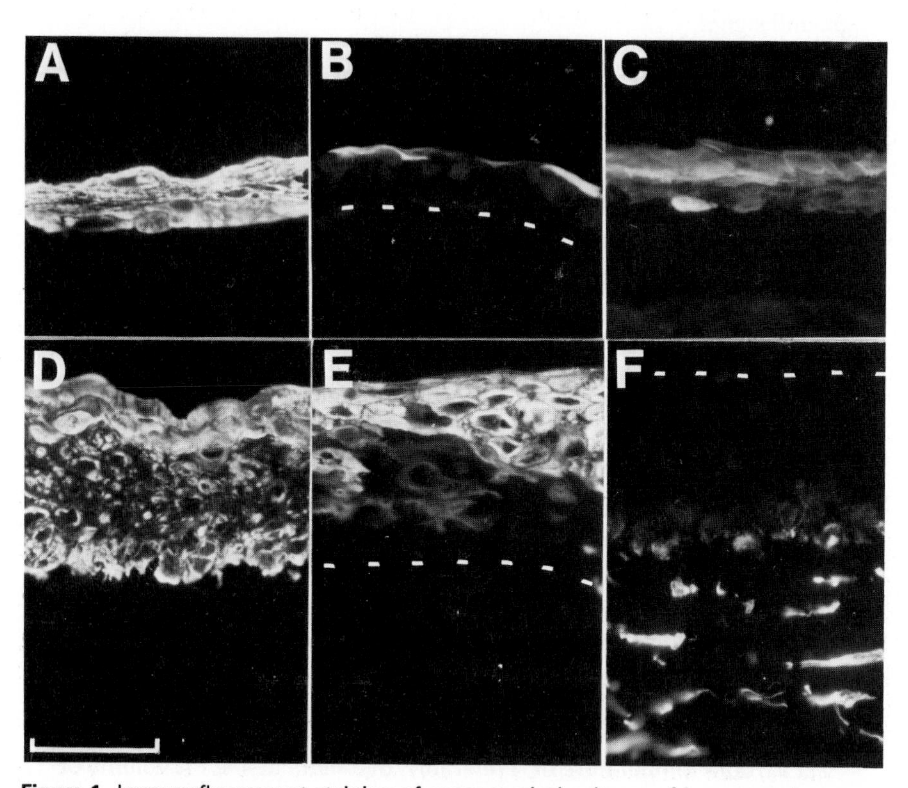

Figure 1. Immunofluorescent staining of organotypical cultures of human oral mucosa. (A–C) Epithelial cells grown on collagen gel alone; (D–F) epithelial cells grown on collagen gel containing mucosal fibroblasts. All specimens were grown in DMEM with 10% fetal bovine serum and harvested 10 days after plating the epithelium. Cytostat sections were prepared and processed for immunofluorescent staining as described in *Protocol 1*. The mAb against K14 (A, D) produces generalized epithelial staining and shows the enhanced growth and organization of epithelium in the presence of fibroblasts. Staining with mAb against K13 (B, E), a cytokeratin normally expressed suprabasally as a differentiation marker in oral mucosa, shows little expression in the epithelium grown on collagen alone but fairly normal expression in the presence of fibroblasts (epithelial/collagen interface marked by broken line). Staining of epithelium for K19 (C), a cytokeratin not normally expressed suprabasally in mucosal epithelium *in vivo*, shows suprabasal expression in epithelium grown on collagen alone. Staining with an antibody against vimentin (F) demonstrates the network of fibroblasts present in the collagen gel. Weak staining of the epithelial basal cells for vimentin is also seen (Bar = 50 μm).

3. Electrophoresis of cytokeratins

Electrophoresis has the advantages of allowing identification, and to some extent quantitation, of each of the individual cytokeratins in a given sample. Cytokeratin extracts (*Protocol 2*) can be examined by one- or two-dimensional polyacrylamide gel electrophoresis (PAGE) as described in *Protocols 3* and *4*. One-dimensional PAGE provides only limited information about cytokeratin components unless it is combined with immunoblotting. The use of two-dimensional PAGE to separate cytokeratins by non-equilibrium pH gradient electrophoresis in the first dimension, and non-gradient SDS slab gel electrophoresis in the second dimension, gives effective separation of proteins by both charge and molecular weight. Such preparations permit identification of individual cytokeratins by comparison with the Moll (4) catalogue (see *Figure 2*).

Protocol 2. Extraction of cytoskeletal proteins (according to Moll *et al.* (4))

Equipment and reagents

- Buffer A (100 ml): 11.183 g KCl and 0.5 ml Triton X-100; make up to 100 ml with buffer B
- Buffer B (200 ml): 372 mg EDTA and 139 mg phenylmethylsulfonyl fluoride (solubilized in isopropyl alcohol); make up to 200 ml with 10 mM Tris-HCl buffer (pH 7.4)
- Cold PBS
- 0.25% trypsin or 3 mM EDTA
- Homogenizer
- Centrifuge and tubes
- Vortex mixer
- Rubber policeman
- Protein Assay Standard II (Bio-Rad)
- Urea solubilization buffer (25 ml): mix 14.25 g urea, 390 mg DTT, 1.25 ml Ampholine (Pharmacia, pH range 3.5–10), 5 ml 10% NP-40, 5 mg leupeptin, and 63 mg SDS; make up to 25 ml with double-distilled water. Aliquot into 1 ml tubes and store at −70 °C

A. *Preparation of cytokeratin extracts from compound tissues*
(From organotypic cultures and control tissues such as oral mucosa or skin)

1. Wash the tissue with cold PBS, cut it into 1 mm^2 pieces, and incubate in: (a) 0.25% trypsin at 4 °C for 5–6 h, **or** (b) 3 mM EDTA at 37 °C, with shaking, for 2–3 h.
2. Remove the epithelial sheets from the connective tissue with fine forceps.
3. Wash the cell sheets twice with cold PBS.
4. Homogenize the cells in cold high-salt buffer (buffer A) and transfer to a centrifuge tube.

B. *Preparation of cytokeratin extracts from epithelial cells growing on plastic*

1. Wash the cells twice with 10 ml of cold PBS.
2. Add 1.5 ml cold buffer A to cover the cells and leave for 2–3 min.
3. Scrape the cells off the bottom of the flask with a rubber policeman and transfer to a centrifuge tube.

Protocol 2. *Continued*

C. *Extraction of cytoskeletal proteins*

1. Take the tubes from part A, step 4 or part B, step 3 and centrifuge (12 000 *g* for 5 min at 4 °C). Remove the supernatant, add 1 ml of fresh cold buffer A to the pellet, vortex and place at 4 °C for 20 min.
2. Centrifuge the sample (12 000 *g* for 10 min at 4 °C) and remove the supernatant.
3. Wash the pellet twice in 0.5 ml low-salt buffer (buffer B) at 4 °C for 5 min each time, and remove the supernatant.
4. Briefly wash the pellet with 0.5 ml cold PBS, centrifuge it again and remove the supernatant.
5. Add 100 μl of urea solubilization buffer to the pellet, vortex and leave at room temperature for 1 h.
6. Centrifuge the sample (14 000 *g* for 15 min at 4 °C), and collect the supernatant which contains the solubilized cytokeratin extract.
7. Assay for protein content (e.g. with Protein Assay Standard II (Bio-Rad) following the manufacturer's instructions), and pipette aliquots containing about 25 μg of total protein (enough for 1 gel) into small tubes. Either use immediately for electrophoresis or store at –70 °C.

Protocol 3. Non-equilibrium pH gradient electrophoresis (NEPHGE) of cytokeratins (according to O'Farrell *et al.* (15))

Reagents

- Acrylamide solution: 28.38 g acrylamide and 1.62 g bisacrylamide made up to 100 ml with double-distilled H_2O. Filter through a Whatman No. 1 paper. Store at 4°C in the dark
- Overlay solution: mix 648 mg urea, 60 μl Ampholine (Pharmacia, pH range 3.5–10) and 642 μl double-distilled water
- Cathode buffer: 20 mM NaOH
- Anode buffer: 17 mM H_3PO_4
- Internal standards: e.g. 10 μl of solution containing 5 μg BSA, 2 μg 3-phosphoglycerokinase and 10 μg rabbit α-actin
- 10 % NP–40
- Equilibrium buffer (200 ml): 3 g Tris base (pH with 1 M HCl to 6.8), 4 g SDS, 20 ml glycerol make up to 200 ml with double-distilled water (pH 6.8)
- Hoefer tube gel unit or equivalent

Method

1. Prepare the tubes (12.5 cm long, 3 mm internal diameter) for tube gels by washing in sulphuric acid, rinsing in double-distilled water, and drying in air.
2. Prepare the NEPHGE gel in a glass beaker, with a magnetic stirrer, according to the number of tube gels needed:

	1.25 ml (1 tube)	5 ml (4 tubes)	10 ml (8 tubes)
Urea	687 mg	2.75 g	5.5 g
Acrylamide/ bisacrylamide	166 μl	0.66 ml	1.33 ml
NP-40 (10%)	250 μl	1.0 ml	2.0 ml
Ampholine (pH 3–10)	62.5 μl	0.25 ml	0.5 ml
Double-distilled water	246 μl	0.99 ml	1.97 ml

De-gas for 2–3 minutes in a dissecting jar connected to a vacuum line. Then add:

Ammonium persulfate (10%)	1.25 μl	5 μl	10 μl
TEMED	0.88 μl	3.5 μl	7 μl

3. Seal one end of the gel tube with Parafilm and use a fine glass pipette to transfer the gel solution into the tube to about 20 mm from the top.
4. Overlay the gel with double-distilled water and stand the tubes vertically in a levelled rack. After 2 h, remove the double-distilled water overlay.
5. Add the cathode buffer into the lower tank and insert the gel tubes into the rubber gaskets. At this and all other stages, avoid creating bubbles!
6. Using a thin pipette tip, add about 25 μg of cytokeratin extract (see *Protocol 2C*, step 7) to the top of the gel.
7. Add the internal standards.
8. Overlay with 25–50 μl of overlay solution.
9. Add the anode buffer to the upper tank and connect the power supply (positive terminal to the top tank). Run at 400 V for 5 h (2000 Vh).[c]
10. Disconnect the power supply and remove the tubes. Release the gels from the tubes by carefully passing a thin needle on a syringe a short way between the gel and the wall of the tube while gently squirting double-distilled water to extrude the gel.
11. Place the gel into equilibrium buffer for 30 min noting its orientation. The gel is now ready for electrophoresis in the second dimension or it can be stored at – 20 °C in a sealed glass tube.

[a] Unless otherwise stated, all reagents are from Bio-Rad.
[b] Acrylamides are toxic so handle with appropriate care.
[c] Note that for NEPHGE the arrangement of the buffers and the terminal connections are *reversed* from the configuration of the apparatus used for iso-electric focusing.

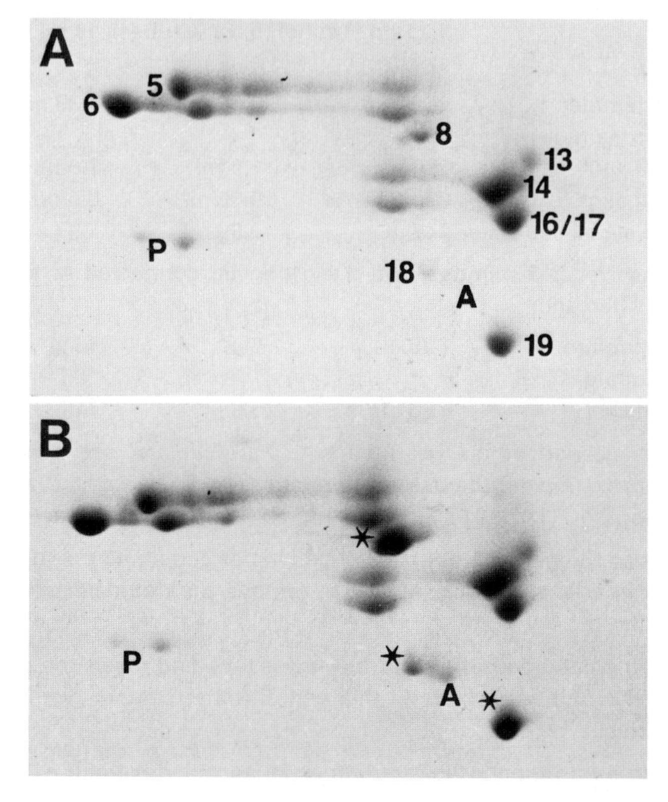

Figure 2. Two-dimensional SDS PAGE gels of cytokeratin extracts from human buccal epithelial cells grown in (A) DMEM with delipidized FBS and (B) in the same medium supplemented with 1 μM retinoic acid. In panel A, the major keratins are labelled numerically according to the Moll (4) catalogue. These are: K5 and K14, the cytokeratin pair expressed by basal cells of stratifying epithelia, K6 and K16, cytokeratins expressed normally by some mucosal epithelia and by hyperplastic epidermis, K8 and K18, cytokeratins usually expressed by simple but not stratifying epithelia *in vivo*, and K19, a transitional keratin expressed by simple epithelia and by some mucosal epithelia. Weak expression of K8, K18, and K19 is seen in the control culture (A) and is greatly enhanced (asterisks) by retinoic acid (B). Standards are 3-phosphoglycerokinase (P) and rabbit α-actin (A).

Protocol 4. Second dimensional (slab gel) electrophoresis[a]

Reagents

- SDS running gel: 10 ml acrylamide: bisacrylamide (29.2:0.8), 11.25 ml 1 M Tris buffer (pH 8.8), 0.15 ml 20% (w/v) SDS, 8.5 ml double-distilled water, 10 μl TEMED, 66 μl 10% alkaline phosphatase (AP)
- 1% (w/v) Bromophenol Blue
- Storing solution: 10% (v/v) acetic acid in 25% (v/v) methanol
- SDS stacking gel: 1.33 ml acrylamide: bisacrylamide (29.2:0.8), 1.25 ml 1 M Tris-HCl buffer (pH 8.8), 50 μl 20% (w/v) SDS, 7.3 ml double-distilled water, 5 μl TEMED, 50 μl 10% AP
- Destaining solution: 10% (v/v) acetic acid in 50% (v/v) methanol
- Acetone:chloroform (1:1)

- SDS running buffer (10 × concentrated stock solution): 30 g Tris, 144 g glycine, 10 g SDS, make up to 1 litre with double-distilled water. To make a working solution, add 9 parts of double-distilled water to 1 part concentrated buffer. Store at 4°C to chill for several hours prior to use
- Hoefer vertical slab unit and 3 mm tube gels
- 0.125% (w/v) Coomassie Blue stain: mix 625 mg Coomassie Brilliant Blue R250, 200 ml methanol, 200 ml double-distilled water, and 50 ml glacial acetic acid
- 1% agarose: mix 1 g agarose, 6.25 ml 1 M Tris buffer (pH 6.8), and 0.5 ml 20% (w/v) SDS, water to 100 ml, heated and then cooled to about 70°C

Method

1. Clean detergent pre-washed glass plates with acetone:chloroform (1:1). Use 0.5 mm spacers if gels are to be used for Western blotting and 0.75 mm spacers for Coomassie staining. Set up the assembly on the stand.
2. Transfer the running (lower) gel solution into the plates. Avoid creating bubbles by pouring the gel from a beaker into a large syringe and running it down the spacer at one side. Fill the plates with gel solution to about 25–30 mm from the top. Overlay with double-distilled water and allow to set for 1 h.
3. Remove the overlay and wash the gel surface with a small amount of double-distilled water. Layer the stacking (upper) gel on to the running gel so that it almost fills the plates. Overlay with a small amount of double-distilled water and allow it to set for 1–2 h.[b]
4. Add the running buffer to the lower tank. Assemble the plates to the top tank and insert into the lower tank.
5. Place the equilibrated NEPHGE gel (see *Protocol 3*, step 11) into the slot of the top tank taking care not to stretch or squeeze it. Ensure that the gel is completely defrosted if it has been stored frozen. Pour melted 1% agarose into the space between the tube gel and the top of the slab gel and partially cover the tube gel. Allow it to set for 15 min.
6. Add running buffer to the top tank and add 10 μl of 1% Bromophenol Blue to it. Connect the unit to a power supply (cathode to the top tank and anode to the lower) and run at 30 mA until the dye front just moves off the gel (about 4 h). Use chilled running buffer or a refrigerated circulating apparatus to avoid overheating the gel.
7. Disconnect the power supply and separate the plates. Discard the stacking gel and stain the lower gel in 0.125% Coomassie Brilliant Blue on an orbital shaker overnight.
8. Place the gel in the destaining solution on an orbital shaker and change the solution hourly until an acceptable level of background is achieved.[c] Store the gel in the storing solution.

[a] Depending on the thickness of the tube gels and the design of the electrophoretic apparatus used, the details of the assembly for electrophoresis will vary. The description here is for a Hoefer vertical slab unit and 3 mm tube gels.

Protocol 4. *Continued*

[b] 1D SDS PAGE. Prepare the apparatus and the running and stacking gels as described in steps 1–3. Form sample wells by inserting a comb into the stacking gel from the top of the plates. After extraction of the sample in buffer B, solubilize it in SDS sample buffer (2.3% (w/v) SDS, 5% (v/v) β-mercaptoethanol, 62.5 mM Tris buffer (pH 6.9) and 10% (v/v) glycerol) and heat it to 100 °C for 10 min. Cool the sample on ice and spin it down (12 000 g for 10 min). Take the supernatant and add 1% Bromphenol Blue to it. If the sample was previously solubilized in urea buffer, add an equal volume of double concentration SDS sample buffer before loading. Run the gel as described above.

[c] Photograph the gel using a yellow or orange filter.

4. Western blotting of cytokeratins

Cytokeratin polypeptides can be transferred to a solid membrane after either one- or two-dimensional separation (see *Protocol 5*). Such transfers are used either for (a) immunoblotting with specific antibodies to identify particular keratins or (b) determining the specificity of antibodies by their observed binding to known keratins (see *Protocol 6*). Stained transfers may also be used to detect the presence of cytokeratins more sensitively than Coomassie stained gels.

Protocol 5. Electrophoretic transfer of proteins[a]

Equipment and reagents

- Ponceau S 10 × stock solution: mix 2 g Ponceau S, 30 g TCA 30 g sulfosalicylic acid in double-distilled water and make up to 100 ml. Working solution: 1 part stock:9 parts double-distilled water
- Tris-buffered saline (TBS): 0.9% (w/v) NaCl in 10 mM Tris–HCl (pH 7.2)
- Transfer buffer: mix 14.5 g Tris, 66.4 g glycine, 1200 ml methanol, 3 g SDS, make up to 6 litres with double-distilled water. Cool to 4°C prior to use
- Hoefer TE-42 transfer electrophoresis unit
- Nitrocellulose transfer membrane (0.45 mm pore size; Hoefer)

Method

1. Cut four sheets of filter paper (Hoefer's TE-42 equivalent to Whatman No. 1) and a sheet of transfer membrane to a size a little larger than the 2D gel. Wear gloves to avoid any contamination of the transfer membrane with lipid or protein. Soak the filter paper and the support pads in the transfer buffer. Wet the transfer membrane from underneath by floating on the transfer buffer for about 5 min, then immerse it.
2. Make a 'sandwich' of, in sequence: a foam support pad, two sheets of filter paper, the transfer membrane, the 2D gel, two sheets of filter paper, and the other foam pad. Take care to keep the assembly wet and to exclude all air bubbles, particularly between the membrane and the gel.

3. Place the 'sandwich' into the plastic cassette and assemble the electrophoresis chamber with the transfer membrane positioned on the anode (positive) side of the gel.
4. Carry out the electrophoresis overnight at 15 V, or for 2.5–3 h at 50 V. Control overheating of the gel by using precooled buffer and, if available and necessary, a recirculating cooling unit.
5. Disconnect the power, remove the gel, and transfer it into a tray containing TBS. Notch one corner of the transfer for orientation and remove gel debris by gentle shaking. The membrane is now ready for immunoblotting and/or staining.

[a] A range of transfer units is commercially available. The Hoefer TE-42 transfer electrophoresis unit is satisfactory and the basic method used is according to Towbin *et al.* (16).

5. Staining transfers for total protein

The transferred proteins can be stained:

(a) to assess the technical success of transfer;

(b) as a more sensitive detection method for cytokeratins than can be achieved with Coomassie staining of gels; or

(c) to identify proteins of interest for immunostaining.

Some staining methods interfere with subsequent immunoblotting and, although a duplicate transfer could be stained, it is usually more convenient to use a single transfer, either staining with a method compatible with subsequent immunostaining, or staining after immunolocalization of proteins of interest.

Ponceau S is a water-soluble dye suitable for the temporary staining of transferred proteins prior to immunostaining. It is not a strong stain but can locate standards and major polypeptides. These can be marked for subsequent orientation by making small pinholes in the membrane before washing in double-distilled water to remove the dye.

For more sensitive staining, use India ink (Higgins India ink or Pelikan Fount India drawing ink, 100 µl in 100 ml of 0.4% (v/v) Tween-20 in PBS). Block the membrane in two changes of 0.4% Tween-20 in PBS and immerse with gentle shaking for 30–60 min in the ink solution. Wash in PBS until background staining is removed. The membrane can be kept between two Whatman papers after drying. Immunoblots can also be stained with India ink but not until after the antibody binding has been displayed (see *Figure 3*). If this method is to be used, block the unbound sites of the membrane with 0.4% Tween-20 in PBS prior to starting *Protocol 6*.

Protocol 6. Immunoblotting with anti-cytokeratin antibodies

Equipment and reagents

- TBS (see *Protocol 5*)
- TBS with 0.5% (w/v) BSA
- PBS
- Substrate working solution: 5 mg naphthol AS BI phosphoric acid (sodium salt), 2 drops DMF, 10 ml veronal acetate buffer, and 5 mg Fast Red TR salt; mix well and filter through a 0.4 μm filter before use
- 2% (w/v) non-fat dry milk (e.g. Carnation natural non-fat dry milk) in PBS with 0.02% sodium azide
- Veronal acetate buffer: 0.9715 g sodium acetate (trihydrate), 1.4715 g sodium barbitone, 247.5 ml double-distilled water, and 2.5 ml 0.1 M HCl; adjust pH to 9.2

Method

1. Rinse the transfer membrane with TBS.
2. Block non-specific absorption of antibodies to the membrane by incubating with 2% non-fat dry milk at room temperature for 5 h (or at 4 °C overnight) with gentle agitation.
3. Wash twice for 5 min each time with TBS/0.5% BSA.
4. Incubate the membrane with the primary antibody at room temperature for 2 h with gentle shaking.[a]
5. Wash three times for 5 min each time with TBS/0.5% BSA.
6. Incubate the membrane in the secondary antibody (alkaline phosphatase or peroxidase conjugated; DAKO) at room temperature for 2 h with gentle shaking.[b]
7. Wash the membrane three times with a total of 500 ml of TBS/0.5% BSA, then rinse the membrane once with TBS solution. It is now ready for detection of Ab binding.[c]

[a] Optimal dilution of the antibody should be predetermined; as a guide use a 1:5 dilution of supernatants from mAb hybridomas and greater dilutions of polyclonal Abs or commercial mAbs.

[b] Typically, a 1:100 dilution of the secondary Ab in TBS/0.5% BSA is effective.

[c] *For alkaline phosphatase conjugated antibodies.* Incubate the membrane in veronal acetate buffer for 5 min, then incubate in alkaline phosphatase substrate working solution for 10–20 min. Stop the reaction by rinsing the membrane with 20 mM EDTA (pH 8.0). Use a green filter for photography.

For peroxidase conjugated antibodies. Use TMB substrate kit (Vector) according to the manufacturer's instructions. Use a yellow or orange filter for photography.

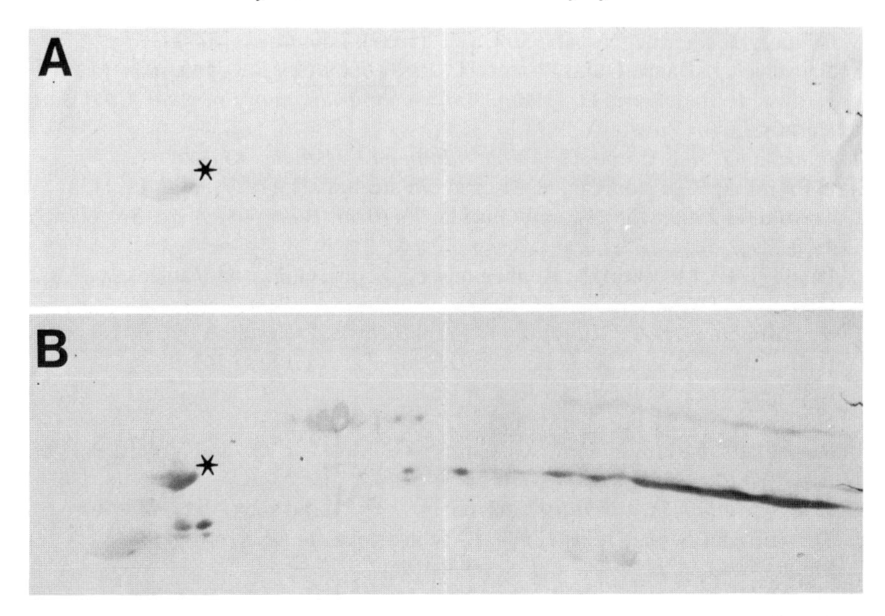

Figure 3. (A) Two-dimensional SDS PAGE gel of mouse cytokeratins transferred to a nylon membrane and immunostained with a mAb reactive with murine cytokeratins (see *Protocols 3–6*). Binding of the primary antibody, which was of unknown specificity, was localized by a peroxidase-conjugated secondary antibody and is apparently localized to a single cytokeratin. (B) Subsequent staining of the immunoblot with India ink localizes the other cytokeratins on the transfer and identifies the reactive cytokeratin as an acidic cytokeratin.

References

1. Dabelsteen, E. and Clausen, H. (1992). *Carbohydrate pathology*. APMIS Supplement 27, Vol. 100. Munksgaard, Copenhagen.
2. Watt, F. M. and Hertle, M. D. (1994). *The keratinocyte handbook* (ed. I. M. Leigh, E. B. Lane, and F. M. Watt), pp. 153–64. Cambridge University Press.
3. Fine, J. (1994). *The keratinocyte handbook* (ed. I. M. Leigh, E. B. Lane, and F. M. Watt), pp. 181–99. Cambridge University Press.
4. Moll, R., Franke, W. W., Schiller, D. L., Geiger, B., and Krepler, R. (1982). *Cell*, **31**, 11.
5. Purkiss, P. E., Steel, J. B., Mackenzie, I. C., Nathrath, W. B. J., Leigh, I. M., and Lane, E. B. (1990). *J. Cell Sci.*, **97**, 39.
6. Morley, S. M. and Lane, E. B. (1994). *The keratinocyte handbook* (ed. I. M. Leigh, E. B. Lane, and F. M. Watt), pp. 293–321. Cambridge University Press.
7. Powell, B. C. and Rogers, G. E. (1994). *The keratinocyte handbook* (ed. I. M. Leigh, E. B. Lane, and F. M. Watt), pp. 401–38. Cambridge University Press.
8. Mackenzie, I. C. (1975). *J. Periodont. Res.* **10**, 49.
9. Sainte-Marie, G. (1962). *J. Histochem. Cytochem.*, **10**, 250.
10. Lane, F. B. and Alexander, C. M. (1990). *Cancer Biol.*, **1**, 165.

11. Cooper, D., Shermer, A., and Sun, T. T. (1985). *Lab. Invest.*, **52**, 243.
12. Schweizer, J., Baust, I. and Winter, H. (1989). *Exp. Cell Res.*, **184**, 193.
13. Harlow, E. and Lane, D. (1988). *Antibodies: a laboratory manual.* Cold Spring Harbor, Labor Press, NY.
14. Johnson, G. and Aranjo, G. (1981). *J. Immunol. Methods*, **43**, 349.
15. O'Farrell, P. Z., Goodman, H. M., and O'Farrell, P. H. (1977). *Cell*, **12**, 1133.
16. Towbin, H., Staehelin, T., and Gordon, J. (1979). *Proc. Natl Acad. Sci. USA*, **76**, 4350.
17. Puts, J. J. G., Moesker, O., Kennemans, P., Vooijs, G. P., and Ramaekers, F. C. S. (1985). *Int. J. Gynecol. Pathol.*, **4**, 300.
18. Van Muijen, G. N. P., Ruiter, D. J., Franke, W. W., Ackstätter, T., Haasmoot, W. H. B., Ponec, M., and Warnaar, S. O. (1986). *Exp. Cell Res.*, **162**, 97.
19. Rodrigues, M., Ben-Zvi, A., Krachmer, J., Schermer, A., and Sun, T.-T. (1987). *Differentiation*, **34**, 60.
20. Ramaekers, F., Huysmanns, A., Schaart, G., Moesker, O., and Vooijs, P. (1987). *Exp. Cell Res.*, **170**, 235.
21. Lane, E. B. (1983). *J. Cell Biol.*, **92**, 665.
22. Troyanovsky, S. M., Guelstein, V. I., Tchipysteva, T. A., Krutovskikh, V. A., and Bannikov, G. A. (1989). *J. Cell Sci.*, **93**, 419.
23. Stasiak, P. C., Purkis, P. E., Leigh, I. M., and Lane, E. B. (1989). *J. Invest. Dermatol.*, **92**, 707.

3

Assessment of cell polarity

ELAINE J. HUGHSON and ROBERT P. HIRT

1. Introduction

1.1 Epithelial cell polarity

The epithelial sheets of metazoa, such as those lining the cavities of the lung, the kidney, or the digestive tract, can be compared to the plasma membrane of unicellular organisms (1). They are the site of complex exchanges between the outside world and the internal milieu which are essential to maintaining a constant body fluid composition. Epithelial cells, therefore, in addition to forming a simple barrier, are the site of numerous vectorial transport processes which are the bases of functions such as water and ion homeostasis, metabolite uptake and secretion (2, 3), and ultimately, protection against pathogens (4).

The functions of epithelial cells are dependent on a specialized cellular organization in which three distinct cell surface compartments can be defined biochemically (i.e. they have distinct protein and lipid compositions) as well as functionally (3). These are:

(a) the apical membrane which faces the external milieu;

(b) the lateral domain where cell–cell contacts take place; and

(c) the basal domain which interacts with the extracellular matrix.

The lateral and basal domains, commonly referred to as the basolateral domain, face the internal milieu and are in contact with the blood supply. To establish and maintain this polarized structure, elaborate intercellular and intracellular organization and complex membrane trafficking pathways are required (2, 3).

Intercellular adhesion is mediated by cell adhesion molecules (5) and the junctional complex (6). The junctional complex, located in the lateral domain, is formed by four structures organized in a distinct order:

(a) the tight junctions are located at the apical/basolateral interface and prevent both the intermixing of apical and basolateral components, and the free diffusion of ions and macromolecules between the lumen and the serosa (7);

(b) the belt desmosomes and

(c) the desmosomes stabilize the epithelial sheet (8);

(d) gap junctions are implicated in intercellular communications, as they couple the cells of the monolayer electrically as well as metabolically (9).

Adhesion to the basement membrane is mediated by receptors to laminin, collagen, and proteoglycans organized into specialized cell–substratum adhesion structures called the hemidesmosomes (8).

The cytoskeleton of epithelial cells shows a particularly complex organization by forming a continuum throughout the epithelium via connections to the junctional complex (10). These interactions are implicated in the stabilization of the epithelial sheet. In addition, the cytoskeleton is known to be involved in sorting and targeting steps in membrane trafficking, including directing vesicular traffic along microtubules to the apical surface, regulating apical endocytosis, and interacting with basolateral membrane proteins in the submembranous cytoskeleton (11).

Membrane proteins with an asymmetric distribution have to be sorted into carrier vesicles and targeted to their specific locations during their biosynthesis (2, 3). According to the epithelial cell type and the protein, two sorting sites have been shown to direct the biosynthetic route of membrane proteins to their respective cell surfaces (12–14). In a kidney-derived cell line (see Section 1.2), intracellular sorting mainly occurs. In contrast, hepatocytes sort membrane proteins in an endosomal compartment following internalization from the sinusoidal membrane (the basolateral membrane); there, apical proteins are sorted into specific carrier vesicles which are subsequently targeted to the opposite membrane (the transcytotic route). Intestinal cells show an intermediate phenotype in that some apical proteins are directly targeted to their final destination while others are transiently expressed at the basolateral membrane. For original papers see references in 12–14.

In the last 20 years or so, continuous progress has been made in elucidating the molecular sorting signals which direct proteins to either one domain or the other (12–14). However, the characterization of the sorting machinery itself is still in its infancy and represents one of the major challenges for modern cell biologists. The development of *in vitro* culture systems on permeable supports was of crucial importance for current research on epithelial cell polarity (15, 16). This was paralleled by the development and application of numerous biochemical and morphological methods to assess the distribution of membrane components underlying that polarity. In the next section we discuss one of the major models used in unravelling the cellular and molecular elements and mechanisms underlying the establishment and maintenance of epithelial cell polarity.

1.2 The MDCK cell line as a model system

MDCK (Madin–Darby canine kidney) cells, derived from the kidney of a Cocker spaniel (17), have been intensively studied by both physiologists and cell biologists and are currently the best characterized epithelial cell model. These cells spontaneously differentiate to form highly polarized monolayers in culture (18–20) and show features of distal kidney tubular epithelia (21). The degree of cell polarity is enhanced when the cells are grown on a permeable support (16), as this more closely resembles the *in vivo* situation where cells obtain nutrients and hormonal signals through their basolateral surface. In addition, such culture conditions give experimental access to both apical and basolateral surfaces for biochemical methods and make possible the study of transepithelial transport processes (15). Two strains of MDCK cells have been characterized (18), and whilst both develop high surface polarity, they exhibit several differences which are summarized in *Table 1*. Although the relatively higher tightness of MDCK I monolayers, as indicated by higher transepithelial resistance measurements (see *Table 1* and Section 3), is attractive for many applications, MDCK II cells appear to be easier to maintain in culture and to manipulate for transfections, and thus have tended to be preferred in numerous studies (25).

The aim of this chapter, using the MDCK cell line for illustration, is to present various methods for assessing epithelial cell surface polarity. We shall first describe general protocols for the maintenance and culture of these cells,

Table 1. Comparison of MDCK I and MDCK II cell lines

	MDCK I	MDCK II	Reference
Similarities			
Same density of tight junctions			(22)
Same amount of the tight junction associated protein ZO-1			
Develop high surface polarity when grown on porous support			(20)
Similarity of polarity for transferrin receptor			(23)
Differences			
Transepithelial electrical resistance (ohm × cm^2)	≥ 1000	100–500	(16)
Morphology of cells grown on porous support		taller and longer microvilli	(18)
Expression of Forssmann antigen	–	+	(24)
Adenylate-cyclase sensitivity to adrenalin, vasopressin, and prostaglandin E_1	+	–	(18)
Expression of alkaline phosphatase	–	+	(18)

followed by morphological and biochemical methods to study the polarized distribution of membrane proteins in epithelial cells grown on permeable supports.

2. Cell culture

The key to working with all cell lines is a basic understanding and knowledge of the culture requirements and properties of the cell under investigation. With epithelial cells, some relevant questions to bear in mind include:

- What is the morphology of the cells when grown on plastic or permeable filters?
- What is the range of transepithelial electrical resistance expected when the cells are grown on filters?
- Up to what passage can a given cell line be used?

Armed with such information, one can design proper experimental conditions which can be reproduced. We shall describe the culture conditions for the MDCK I cell line and different ways to assess the quality of the cell monolayer when grown on permeable support. Additional information on *in vitro* cell line culture can be found in ref. 26, and we recommend browsing for background information through such a text to all beginners.

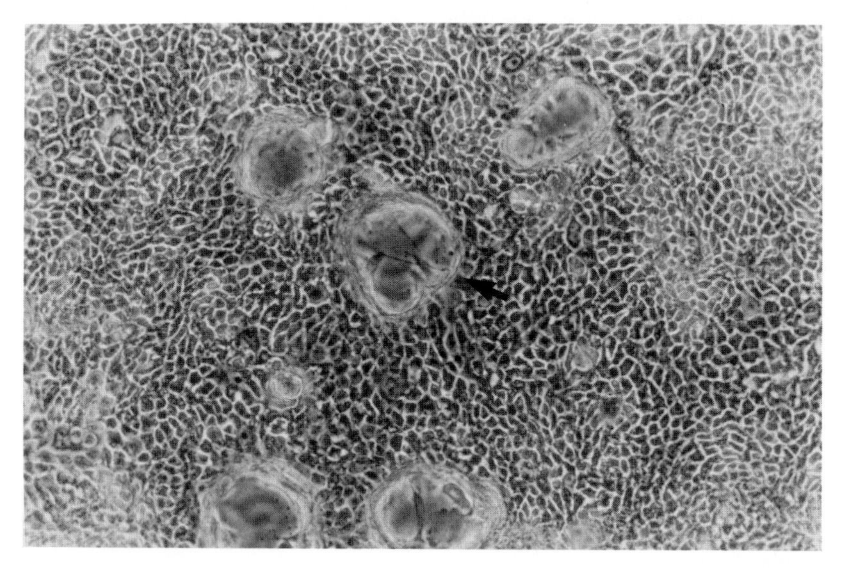

Figure 1. MDCK cells grown in plastic flasks for five days. Note the uniformity of the monolayer, except where 'blisters' have formed (arrow).

2.1 MDCK cell culture on plastic

Stocks of cultured cells are maintained in culture flasks or dishes. Since epithelial cells display well-formed intercellular junctions and strong attachments to the substratum, fairly harsh conditions are needed to passage these cells. *Protocol 1* describes the method used to detach MDCK cells from plastic flasks. The cells can be transferred either to new flasks for expansion, or plated on to a given substrate for experimental work as described in Section 2.2. After 3–5 days in culture on solid support, MDCK cells should be confluent and develop blisters (15) (*Figure 1*). The reason for blister formation is that the Na/K-ATPase in these cells is basolaterally located, and pumps ions downwards, drawing water with them. Because the tight junctions now form a barrier across intercellular spaces, fluid accumulates under the monolayer and, in places, forces it upwards. For long-term storage of cells, frozen stocks in liquid nitrogen are made (see *Protocol 1B*).

Protocol 1. Plating MDCK cells in flasks and preparation of frozen stocks[a]

Equipment and reagents

- Culture medium: Dulbecco's modified Eagle's medium (DMEM) complemented with 10% fetal calf serum (FCS), 2 mM L-glutamine, 10 mM Hepes (pH 7.4), and antibiotics (50 IU/ml penicillin and 50 μg/ml streptomycin) (all from Gibco-BRL)
- 0.25% trypsin–0.2% EDTA solution (Gibco–BRL)
- Phosphate-buffered saline without Ca^{2+} and Mg^{2+} (PBSA) (Gibco–BRL): 10 × stock solution; 1.368 M NaCl, 81 mM Na_2HPO_4, 26.8 mM KCl, 14.6 mM KH_2PO_4; in double-distilled water. Make up to the working dilution with double-distilled water and adjust to pH 7.4. Sterilize by autoclaving or filtration through a 0.22 μm filter. Store at room temperature
- Freezing medium: 45% culture medium, 45% FCS, 10% DMSO
- One flask (typically 25 cm^2 or 75 cm^2) of confluent MDCK I cells
- a 37 °C incubator with a humid atmosphere equilibrated to 5% CO_2
- a 37 °C waterbath
- Sterile tissue culture cabinet
- Sterile plastic or glass pipettes, sterile Pasteur pipettes and vacuum line
- Inverted microscope
- Five new flasks, and a sterile 50 ml tube (Falcon)
- Cryotubes (Nunc)
- Bench centrifuge with variable speed

A. *Cell culture on plastic*

1. Suck off the spent medium with the vacuum line.
2. Wash the cells twice with sterile PBSA and add 1 or 3 ml of the trypsin solution for 25 cm^2 and 75 cm^2 flasks, respectively. Re-incubate at 37 °C.
3. Examine the cells every 5–10 min under an inverted microscope until they are beginning to round up and to detach from the substrate.
4. Add 9 ml of prewarmed culture medium (the presence of FCS will inhibit the trypsin) and gently wash around the flask to remove the cells.
5. Transfer the cells to a sterile 50 ml Falcon tube and centrifuge the cells at 500 *g* for 5 min.

Protocol 1. *Continued*

6. Remove the supernatant from the centrifuged cells (with the vacuum line) and gently resuspend the pellet in 25 ml of culture medium. Add 5 ml of the cell suspension to each flask and replace them in the incubator. The cells should be passaged twice a week. After 2–3 days, the cells are confluent and have a 'cobble-stone' appearance, and should be fairly evenly sized (see *Figure 1*).[c]

B. *Preparation of frozen stock*

1. Trypsinize the cells as indicated in part A (up to step 5).
2. Resuspend the cells in ice-cold freezing medium. Since DMSO is toxic, carry out all steps as quickly as possible.
3. Transfer the cells to cryotubes (e.g. one-fifth of the cells from a flask per tube) and place the tubes at –70°C overnight in a closed polystyrene box to slow down the freezing process.[b]
4. Transfer the tubes rapidly from the –70°C freezer to the liquid nitrogen tank (gloves and glasses must be worn for protection when liquid nitrogen is handled). The cells can be stored for years under these conditions.
5. To thaw the cells, transfer a cryotube directly from the liquid nitrogen to a 37°C waterbath and when thawed, dilute the cells in 10 ml culture medium. Spin the cells at 500 *g* for 5 min, resuspend in fresh culture medium, and plate the cells in a flask. If the same flask size is used, the cells should be confluent in 2–3 days. Remember 'slow freeze, quick thaw' for good results.

[a] Since MDCK I cells do not have a stable phenotype (18), a large stock at low passage (e.g. 50 tubes) should always be prepared at the beginning of a project (e.g. when a new stock of cells has been received or when transfected cell lines have been made). MDCK I cells are used up to passage 20 (16, 18), MDCK II up to passage 40 (27). In general, clonal variations should always be avoided by limiting the number of passages. One passage is defined here as a trypsinization step.
[b] The cells can be stored at –70 °C for up to one month.
[c] Cells can also be plated on coverslips for immunofluoresence studies (see Section 4.1). Coverslips are autoclaved for use, or dipped in ethanol and flamed. Square coverslips (22 mm^2) are placed in 6-well dishes. At Step 6, resuspend the cells in 30 ml medium and add 2 ml of cell suspension to each well (if many cell clones/lines or antibodies are to be tested, round 13 mm coverslips can be placed in 24-well plates).

2.2 Culturing epithelial cells on permeable supports

In addition to the higher level of cellular polarity obtained when cells are cultured on filters (*Figure 2*) (15, 16), these culture conditions have the advantage that transepithelial transport processes can be studied and experimental manipulations can take place at either surface (Section 5). *Protocol 2* describes the plating of MDCK cells on permeable Transwell filters.

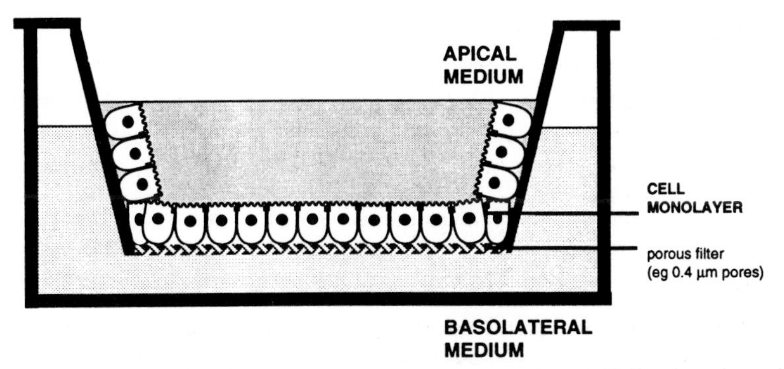

Figure 2. Diagram of MDCK cells cultured on permeable filters. Following the required culture period, a cell monolayer will form and gradually become polarized. Also shown are the cells which tend to grow on the border of the chamber, and for which the state of polarity is different from the ones in direct contact with filter (see Section 5).

Protocol 2. Plating MDCK cells on filters

Equipment and reagents

- Equipment and reagents from *Protocol 1*
- Haemocytometer
- Commercial inserts and multiwell plates (see *Table 2*), e.g. 6 × 24 mm standard Transwells

Method

1. Proceed as above to step 5 as in *Protocol 1*, Part A. Count the cells.
2. Remove the supernatant from the cell pellet, add sufficient culture medium to obtain a dilution of 0.5×10^6 cells/ml, and gently resuspend the cells.
3. Add 2 ml of the cell suspension to the upper chamber of each Transwell. Gently remix the cell suspension between each Transwell. Add 2.5 ml culture medium to the lower chamber.
4. Feed the filters every 1–2 days. In order to avoid hydrostatic pressure on the monolayer from the lower chamber, remove the medium first from the lower chamber and add the medium first to the upper one. The cells should form a tight polarized monolayer after 4–6 days.

There is now a wide choice of filters commercially available with different diameters, pore sizes, and surfaces for growing epithelial cell lines (see *Table 2*). The authors have mainly experience of 24 mm Transwells. If reagents are limited, smaller diameter filters, and hence smaller volumes, may be useful. In our experience, MDCK cells grow best on small (0.4 μm) pore size filters, as they have a tendency to grow through larger sized pores. As these cells

Table 2. Major commercially available filters

Company	**Name**	**Material**	**Diameter (mm)**	**Pore size (μm)**	**Comments**
Costar	Transwell standard		6.5–100	0.1–12	Opaque, 15 μm thick
	Clear	Polycarbonate	6.5–24	0.4–3	Transparent
	Transwell-Col		6.5–24	0.4–3	Coated with bovine collagen I and III
Becton-Dickinson	Falcon insert	Polyethylene-terephtalate	6.3–23	0.45–8	Translucent/ transparent, high pore density available for maximum permeability
	Biocoat		6.4–23	0.45–3	Coated with either bovine collagen I or IV, laminin or fibronectin
Millipore	Millicell[a]				
	-CM	Biopore CM/ polystyrene	12–30	0.4	Transparent, needs coating
	-HA	Cellulose fibres/ polystyrene	12–30	0.45	
	-PCF	Polycarbonate	12	0.45–3	No need for coating

[a] Detailed protocols for manipulations are available from the company.

secrete certain basal lamina components (28, 29), they grow well on uncoated Transwells, but other cell lines may need an exogenous extracellular matrix. Extracellular components are known to influence the level of expression of numerous genes, some of which are involved in the differentiation of epithelial cells (30). Becton–Dickinson offer a good range of coated filters, but filters can also be coated in the laboratory (see Chapter 7, *Protocol 2*).

With translucent filters (*Table 2*) monolayers can be optically monitored and a higher density of cells, compared to confluent plastic flasks, should be observed (20). If the cells are growing well, the medium should become slightly acidified and change colour to orange. As the tight junctions become established, a difference in medium levels between the two chambers should be maintained after feeding. Another simple way to assess the quality of cell monolayers grown on permeable filters is to stain the cells with Toluidine Blue (see *Protocol 3*). This allows the overall morphology of the monolayer to be assessed cheaply and quickly. This protocol could, for example, be useful when numerous new culture conditions are tested. In our experience, high transepithelial resistance of MDCK I cells monolayers was always correlated with homogeneous staining of the monolayer. This approach is particularly useful when opaque filters are used and the morphology of the cells cannot be observed using light microscopy. This stain also reveals the presence of cells on the sides of the chamber supporting the filter up to the air–medium interphase (see *Figure 2*). The relevance of the presence of these cells will be discussed in Section 5. Quantitative means to assess the monolayer tightness are described in the next section.

Protocol 3. Gross assessment of cell monolayer morphology on filters using cytochemical staining[a]

Equipment and reagents

- A filter containing a confluent cell monolayer (*Protocol 2*)
- 70% (v/v) ethanol in double-distilled water
- 1% (w/v) Toluidine Blue (Sigma) plus 1% (w/v) $Na_2B_4O_7$
- PBSA (Gibco–BRL; *Protocol 1*)

Method

1. Rinse both chambers of the filter twice with PBSA.
2. Add 0.5 ml of the Toluidine Blue solution to the cells (upper chamber) and incubate for 30–60 min.
3. Rinse the filter four times with PBSA and four times with 70% ethanol (or until no more dye is removed during the washes).
4. A picture using a Polaroid camera can be taken for the record.

[a] Homogeneous staining is obtained when cells form a 'healthy' monolayer. If some part of the monolayer contains dead cells or aggregates of cells, non-homogeneous staining is observed.

3. Assessment of monolayer integrity

For many experiments, including assessment of cell surface polarity using biochemical techniques (see Section 5), it is essential that the cell monolayer is relatively impermeable. In order to monitor the quality of the epithelial cell monolayer tightness, which depends upon the establishment of functional tight junctions, various approaches have been developed. The principle of these methods is to measure the diffusion through the epithelial monolayer of an enzyme (Section 3.1), a fluorescent dye (see Chapter 6, Section 2.3.3), or a radiolabelled molecule such as mannitol (ref. 31 and Chapter 6, Section 2.3.2). Alternatively, the transepithelial resistance can be measured (Chapter 6, Section 2.3.1).

3.1 Assessment of monolayer integrity using horseradish peroxidase (HRP)

HRP is an extremely active enzyme widely used in molecular and biochemical techniques when conjugated to antibodies and other reagents (e.g. in Western blot, ELISA, *in situ* hybridization). In this instance, the free molecule (of 40 kDa) is used to monitor how leaky the monolayer is. This is a straightforward, non-radioactive method using standard laboratory equipment (see *Protocol 4*). With transfected MDCK I cells expressing the rabbit polymeric immunoglobulin receptor (pIg-R) described in Hirt *et al.* (32), less than 0.05% HRP passed through the monolayer after an overnight incubation (E. Hughson, personal observation).

Protocol 4. Monitoring cell monolayer tightness using HRP

Equipment and reagents

- Monolayers of filter-grown cells (see *Protocol 2*)
- Culture medium (see *Protocol 1*)
- 100 μg/ml HRP (Type I, Sigma) in complete culture medium. Sterilize by passing through a 0.2 μm filter
- PBSA (see *Protocol 1*)
- PBSA/EDTA: PBSA containing 1 mM EDTA
- Substrate solution: 50 mM sodium phosphate buffer, pH 5.0, 342 μM *o*-dianisidine dihydrochloride (Sigma), 0.003% (v/v) H_2O_2. Store at 4 °C protected from light
- 0.1% sodium azide in PBSA
- 96-well, flat-bottomed ELISA plates (Falcon) and an ELISA plate-reader

Method

1. After the required culture time (see *Protocol 2*), aseptically remove the spent medium from both chambers of the filter-grown cells.
2. Add 2 ml complete medium to the apical chamber and 2.5 ml of 100 μg/ml HRP in complete culture medium to the basal chamber.
3. After an overnight incubation (about 16 h), harvest the medium from the apical chamber.
4. *Optional.* As a positive control for leakage, open up the tight junctions by washing the cells twice with PBSA, and adding 2 ml PBSA/EDTA to the apical chamber and 2.5 ml PBSA/EDTA containing 100 μg/ml HRP to the basal chamber. Replace the cells in the incubator for 30 min, and then harvest the apical medium.[a]
5. Make up HRP standards (1:2 serial dilutions) from 3.9 to 1000 ng/ml in culture medium (if step 4 was included, additional standards must also be prepared, diluted in PBSA/EDTA).
6. Add the standards and samples to the 96-well plate, in duplicate, 50 μl / well. Also include blanks (complete medium and/or PBSA/EDTA, without HRP).

7. Add 100 μl substrate solution per well, and after 2 min, add 50 μl of 0.1% sodium azide to stop the reaction.[b]
8. Read the optical density on a plate-reader at 455 nm, and calculate what percentage of HRP has passed through the monolayer. There should be a large increase on addition of EDTA.

[a] Other positive controls for leakage can be empty filters or filters plated with confluent non-epithelial cells (e.g. fibroblasts). Alternatively, MDCK cells can be grown overnight in calcium-free culture medium, conditions in which the tight junctions will open.
[b] If the colour change is too rapid, or the optical densities of the samples do not fall within the range of the standards, it may be necessary to dilute the samples and repeat the assay.

3.2 Measurement of transepithelial electrical resistance

Transepithelial electrical resistance (TER) is routinely measured to monitor electrophysiological properties of epithelial cells (15) and/or to assess monolayer tightness (16) as described in Chapter 6, Section 2.3.1; MDCK cell lines I and II show different electrophysiological properties (15) and develop different TER (*Table 1*) (16).

3.3 Tightness and polarity

It is important to distinguish between tightness of the monolayer and the polarity state reached by the cells in the monolayer. A tight monolayer does not necessarily equate with polarized cells. Morphological (Section 4) or biochemical (Section 5) data are needed to assess properly the polarity state of the cells in the monolayer. For instance, in MDCK cells transformed with the oncogene Ki-*ras*, the polarized distribution of basolateral markers and functional tight junctions was maintained, but the polarized distribution of apical markers was lost (33). Similarly, there is no relation between the level of TER and the level of polarity of the cells as shown for the two MDCK cell lines (*Table 1*). They develop different levels of resistance but both can form highly polarized cell monolayers as judged from functional and morphological data (18, 23). Tight junctions are necessary but are not sufficient for the development of a polarized monolayer. Thus additional properties have to be tested in order to assess the overall quality of the *in vitro* polarized system used. The polarity state can be monitored by comparing some properties in both domains, such as: uptake of amino acids (methionine) (20), polarized secretion of an endogenous protein such as an 80 kDa protein in MDCK II cells (34), or by assaying the rate (efficiency) of ligand uptake by a specific receptor found in one domain (see *Protocol 7* in Section 5) (23).

4. Morphological methods to assess cell polarity

Morphological and biochemical methods used to assess the surface polarity of a cell line are complementary for the information they yield. For biochemical techniques, the monolayer should be fairly tight. These types of procedures give an average, quantitative value for the cell population tested, whereas morphological methods give primarily qualitative data. If quantitative results are to be derived from morphological data, numerous fields must be observed/recorded in order to make any type of quantitative statement (35, 36). Morphological methods can be of particular interest when studying a heterogeneous population. For example, transfected cells are sometimes variable in their expression of exogenous proteins, and it may be useful to know if recloning would be beneficial.

4.1 Fluorescence microscopy

Protocol 5 describes a general method for studying the cell surface polarity of cells grown on coverslips or filters by fluorescence microscopy.

Protocol 5. Investigation of cell surface polarity by indirect immunofluorescence

Equipment and reagents

- Nail varnish
- MDCK cells grown on filters (*Protocol 2*) or coverslips (*Protocol 1*)
- PBSA (*Protocol 1*)
- 3% (v/v) paraformaldehyde (PF)[a] in PBSA (*Protocol 1*). Carry out all manipulations with PF in a fume-hood, using gloves. Use a 10% stock of PF from Polyscience Inc. to prepare this solution
- 50 mM NH_4Cl in PBSA
- 0.2% (w/v) BSA in PBSA
- Permeabilizing solution: either 0.2% BSA/0.05% saponin (Calbiochem) in PBSA or 0.2% Triton X-100 in PBSA[b]
- Blocking solution: 0.2% BSA in PBSA[b]
- Damp-box (airtight box with screwed-up moistened tissue) and Parafilm
- Microscope slides and a fluorescence microscope
- Primary antibody (specific for the protein of interest). Dilute this as necessary in blocking solution[c]
- Secondary antibodies conjugated to a fluorophore such as fluorescein (we generally use DAKO antibodies, at a dilution of about 1: 100 in blocking solution[b])
- Fine forceps
- Vacuum line with a Pasteur pipette
- Mounting medium: 0.3% *n*-propyl gallate in 9:1 glycerol: 10× PBSA. Stir this solution overnight and store at 4°C protected from light.

A. *Light microscopy of cells grown on coverslips*

1. Remove the culture medium, and wash the cells twice in 2 ml PBSA.
2. Fix the cells in 1 ml 3% PF for 20–30 min.
3. Wash the cells twice with PBSA. Do not let the cells dry out.
4. Quench the free aldehyde groups by washing the cells twice with 50 mM NH_4Cl in PBSA for 10 min each time.
5. Permeabilize the cells to gain access to both intracellular and basolateral proteins: this can be done using 0.2% Triton X-100 for 5 min, followed by several washes in PBSA.[b]

6. Place a 50 μl drop of antibody on a piece of *flat* Parafilm in a damp-box. Gently invert a coverslip bearing the cells over the drop, and leave for 30 min.[d]
7. Gently float the coverslip off with some PBSA, replace it in the well (cells uppermost), and thoroughly wash with five changes (2 ml) of blocking solution over 1 h on a gentle shaker.
8. Repeat the incubation procedure (steps 6–7) with the secondary antibody.
9. To mount the coverslip, place a *small* drop (< 50 μl) of mounting medium on a microscope slide. Rinse the coverslip in double-distilled water (it is now critical you know on which side your cells are located!), wipe the back of the coverslip and drain it, then gently lower it on to the drop. Seal the edges with nail varnish.
10. When the nail varnish is dry, examine the cells under the fluorescence microscope.

B. *Light microscopy of filter-grown cells*

1. Remove the medium from both sides of the filter and wash the filter twice with PBSA.
2. Fix the cells and proceed as in part A, steps 2–4.
3. Using a sharp scalpel, carefully cut the filter out of its support. The filter (24 mm diameter) can then be cut into up to eight segments. Since the filter pieces flip over during washing, it is a good idea to cut a small corner off the left-hand side of each piece to allow orientation.
4. Proceed as in part A, steps 5–8.
5. Mount the pieces of filter on microscope slides with coverslips as described in part A, step 9, but unless you are sure which side the cells are on, do not seal the coverslip.
6. When viewing the sample, it is necessary to focus on the filter with the transmission optics, since the cells are not visible.

[a] Some antigens are sensitive to aldehyde fixation, in which case it may be necessary to try acetone or methanol fixation.
[b] A gentler method, which may be more favourable for many antibodies, is to block and permeabilize using BSA/saponin/PBS, 2 × 10 min (saponin must be present in all subsequent blocking, washing, and antibody dilutions, as the permeabilization is reversible).
[c] If the antibody has not been used for immunofluorescence before, try out different dilutions (e.g. 1:20, 1:100, 1:500).
[d] Suitable negative controls are: no primary antibody, use of pre-immune serum (for a polyclonal antibody) or an unrelated antibody of the same species, absorption of antibody with antigen, or for transfected antigens, use non-transfected cells. If possible, also include a positive control, e.g. other cell types known to contain that antigen.

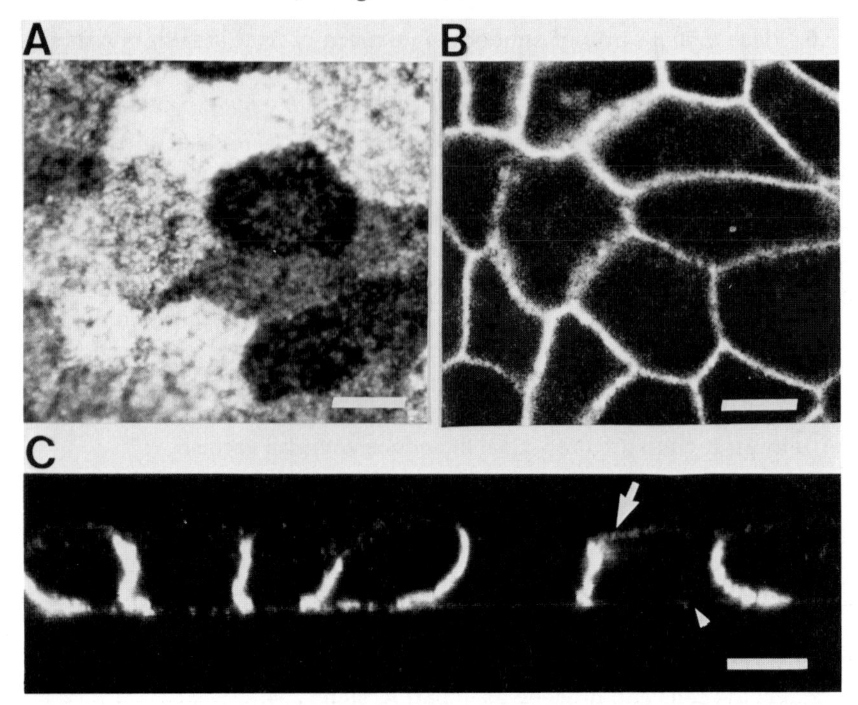

Figure 3. Filter-grown MDCK cells were fixed and tagged with antibodies for CD44, followed by FITC-conjugated secondary antibody. (A) (B) Images were taken using a confocal scanning laser microscope, but similar patterns can be obtained using a conventional fluorescence microscope, by focusing up and down on the monolayer. (A) MDCK cells transfected with a mutant form of human CD44 (labelled with a human-specific antibody) which is expressed only at the apical surface (unpublished data, Drs Humma Sheikh and Clare Isaak, Imperial College, London, UK). The cells show typical stippled apical labelling (due to microvilli). (B) Wild-type MDCK cells labelled for endogenous CD44 (with an antibody recognizing different species). The network pattern is characteristic of lateral staining. Bars, 10 μm. (C) The same specimen as (B), except a vertical section through the monolayer was constructed using a confocal scanning microscope. This image clearly shows that endogenous CD44 is only lateral in distribution (arrow and arrowhead indicate top and bottom of monolayer, respectively). Bar, 10 μm.

Typical apical and lateral labelling are shown in *Figure 3*. Examples of the latter are adhesion proteins such as E-cadherin (37), desmosomal proteins (38), growth factor receptors (39), and transferrin receptors (40). It is necessary to permeabilize the cells even on filters to visualize basolateral proteins, because, following fixation, the intercellular junctions restrict lateral accessibility. Tight-junction antibodies such as anti-ZO-1 (see Chapter 6, Section 2.3.3) show a typical lateral stain but only at the top of the monolayer (22). Aminopeptidases are typical markers for the apical membrane (41).

If no signal is obtained, apart from sensitivity of the antigen to fixation, the species specificity of the antibody should also be considered, *but* some antibodies just do not work for immunofluorescence, even if they work well for biochemistry. If high background fluorescence is a problem, try: diluting the antibody further, additional washes, increasing the concentration of BSA to 1%; or use of other blocking agents, such as gelatin or an appropriate serum may also help.

If access is available to a confocal scanning laser microscope (e.g. Biorad MRC-600), this will be very useful in assessing both surface polarity and character of the culture (cell height and monolayer/multilayer), as vertical sections can be generated (see *Figure 3c*) (for further details see ref. 42). It is possible to obtain similar images without confocal microscopy, but with less resolution and the need for greater technical skill (cutting 0.5–5 μm cryostat sections) (43).

4.2 Electron microscopy

4.2.1 Conventional electron microscopy

Much information may be obtained about the state of polarity attained by cells grown on permeable supports by observing the cells under the electron microscope. An example is the polarized distribution of the various organelles. *Protocol 6* describes a general method to study the overall organization of a cell.

Protocol 6. Processing cells for conventional transmission electron microscopy (TEM)

Equipment and reagents

- Filter-grown MDCK cells (see *Protocol 2*)
- PBSA (see *Protocol 1*)
- 0.1 M sodium cacodylate buffer (Agar Scientific Ltd): use 0.1 M HCl to adjust the pH to pH 7.4. Wear gloves when preparing this solution
- Fixative: 2% PF/2.5% glutaraldehyde (Polysciences) in sodium cacodylate buffer[a]
- 1% (v/v) osmium tetroxide (Taab Laboratories) in distilled water[a]
- 70% (v/v), 95% (v/v), and absolute ethanol (sodium sulfate, pre-dried in an oven, should be added to the stock bottle to dry the alcohol at least 24 h before use)
- Propylene oxide (Agar Scientific Ltd)[a]
- Formvar-coated G100 copper grids (Agar Scientific Ltd) (44)
- Epon (Poly-bed; Polysciences) or similar epoxy resin; we use the following ratio: 1.27 g Epon, 0.78 g DDSA, 0.94 g MNA, and 0.09 g DMP 30. Mix well and avoid air bubbles
- Acetate veronal buffer: 2.43 g sodium acetate and 3.68 g sodium barbitone dissolved in 125 ml double-distilled water
- Kellenberger reagent[b]: 5 ml acetate veronal buffer, 7 ml 0.1 M HCl, 13 ml water, pH 6.0, 0.125 g uranyl acetate (Agar Scientific Ltd), keep in the dark at 4 °C
- Small glass vials
- Fine forceps
- Flat embedding moulds
- Lead citrate (45)
- Disposable scalpel

Method[c]

1. Wash cells twice in PBSA.
2. Fix with PF/glutaraldehyde in sodium cacodylate buffer at room temperature for 30–60 min.

Protocol 6. *Continued*

3. Wash the cells in sodium cacodylate buffer twice, and cut the filter out of its holder with a sharp scalpel.
4. Post-fix the cells in osmium tetroxide for 1 h.
5. Wash the cells thoroughly in double-distilled water. The cells can be left at 4 °C at this stage if required (no longer than overnight).[b]
6. Wash thoroughly in water and, being careful to damage the cells as little as possible, cut the filter into small sections with a scalpel holding the filter with fine forceps.
7. Dehydrate the specimens sequentially through the alcohol series; 2 × 10 min each, and 3 times with absolute ethanol.
8. Transfer the pieces to a glass vial, and incubate in two changes (15 min each) of propylene oxide. The pieces of filter will curl up like a Swiss roll.
9. Incubate the filter pieces for 30 min at room temperature on a rotator with Epon:propylene oxide (1:1).
10. Incubate the pieces with complete Epon twice, for 2 hours each time.
11. Place the specimens in coffin moulds, with labels, for embedding. Orientate the rolls lengthways so that when the block is sectioned, the roll is sliced perpendicular to its length. Be careful not to trim the block too close to the filter or else it will split.
12. Collect sections on Formvar-coated G100 grids to minimize the area of section hidden by grid bars.
13. Stain the sections for about 5 min with lead citrate, in a Petri dish containing sodium hydroxide to absorb carbon dioxide, and wash well in a stream of double-distilled water.

[a] Prepare and use in a fume hood and wear gloves.
[b] *Optional.* To give greater contrast, stain 'en bloc' with Kellenberger reagent for 45 min. Alternatively, use 0.1% uranyl acetate in water for 30 min.
[c] This method is suitable for Costar Transwell inserts.

For the theory and more detailed methodology for electron microscopy, see refs 46 and 47. Typical morphological features of epithelial cells are shown in *Figure 4*. Simple epithelial cells should form essentially a monolayer, with nuclei near the base of the cell. Sometimes the cells may appear multilayer in places, but this may be simply due the intertwining of obliquely leaning cells. The apical surface generally has microvilli, and in the case of intestinal cell lines, such as Caco-2 cells, there is a well-developed brush border. MDCK I cells are fairly tall, with most organelles (including Golgi bodies) in

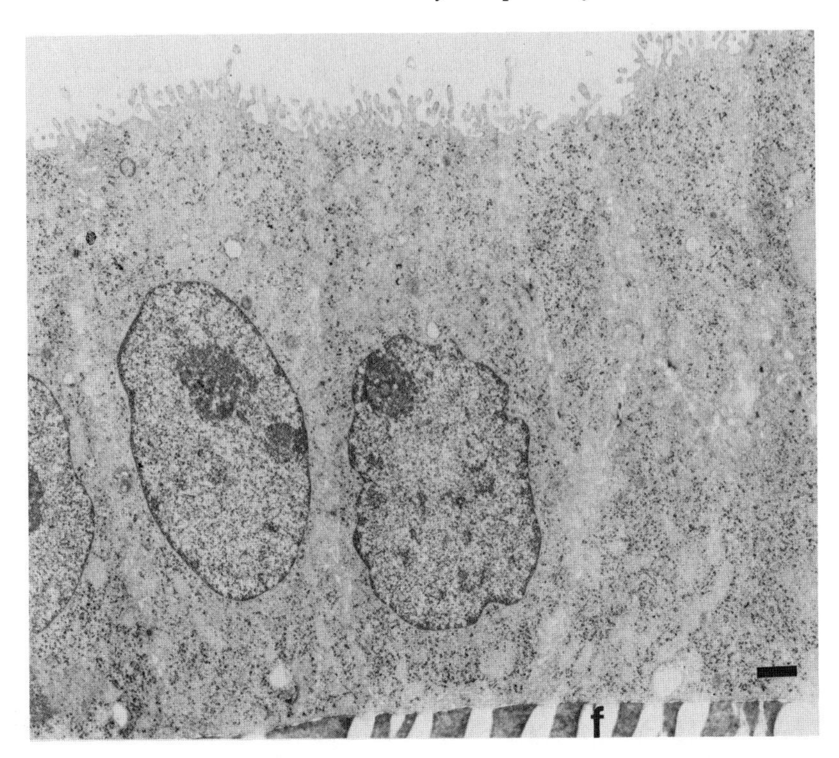

Figure 4. Electron micrograph of a monolayer of filter-grown MDCK I cells. Typical features are basally placed nuclei and apical microvilli. The filter (f) is seen at the base of the cells, the pores of the filter appear as white strips. Bar, 1 μm.

the apical half of the cell (*Figure 5*). The lateral membranes are closely apposed, with apically placed tight junctions and desmosomal complexes scattered down the height of the cell (*Figure 5*).

4.2.2 Immunocytochemical labelling in TEM

The surface polarity of epithelial cells may be studied at the electron microscope level using several methods. Given the difficulties inherent with these techniques, careful consideration should be given as to whether immunofluorescence alone would give sufficient information, especially now that sophisticated light microscopic technology is available. The methods can be grouped according to whether immunolabelling is carried out before or after embedding and sectioning.

i. Pre-embedding methods

Immunoperoxidase (48)

The cells are fixed lightly, permeabilized, immunolabelled directly or indirectly using HRP-conjugated antibodies with diaminobenzidine (DAB) as a

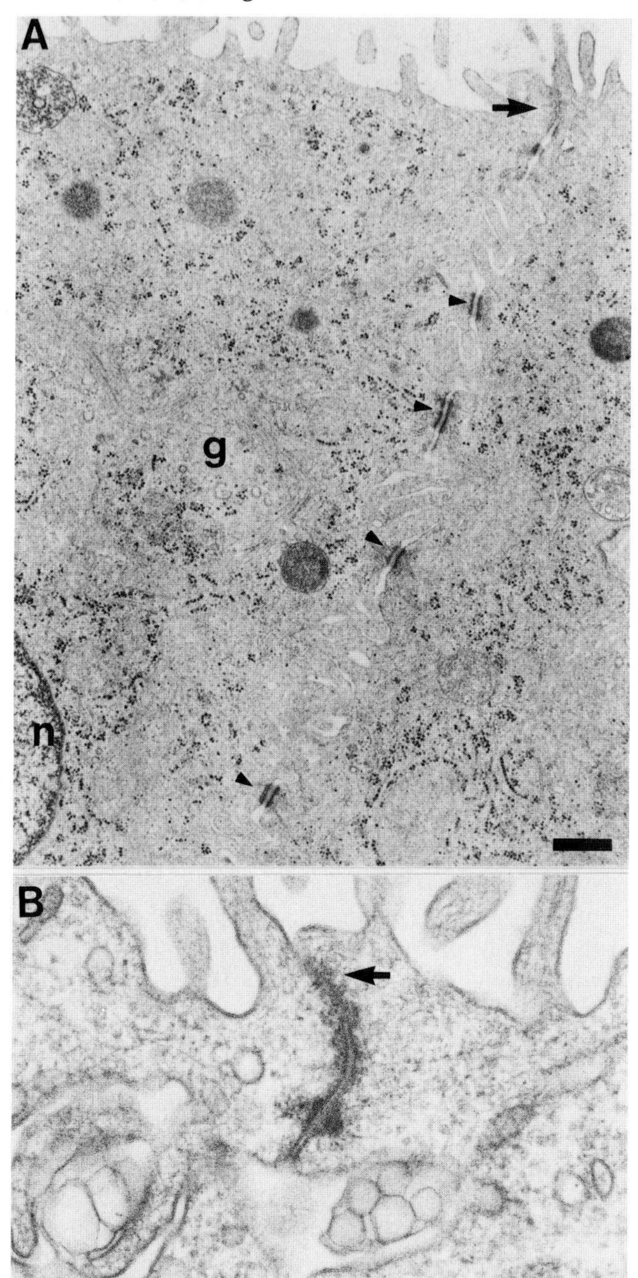

Figure 5. As *Figure 4* at higher magnification showing details of the cellular junctions. (A) Arrowheads indicate desmosomes and arrows, tight junction (n, nucleus). Also note the apically placed Golgi apparatus (g). Bar, 0.5 μm. (B) High magnification showing the apical junctional complexes (arrow, tight junction) (kindly provided by Dr Robert Parton, EMBL, Heidelberg, Germany).

substrate—which gives a black insoluble reaction product—and processed for conventional microscopy (*Protocol 6*). No extra equipment is needed. The method is sensitive, but the DAB reaction product may diffuse during processing; if the permeabilization step is included, morphology is often poor, however, if the permeabilization step is omitted, there may be accessibility problems—see below.

Immunogold

Again, cells are fixed lightly, and labelled with the primary antibody, followed by a secondary antibody conjugated to colloidal gold (available from Biocell Research Laboratories or Amersham Life Sciences), before processing for electron microscopy. The gold particles are easily identifiable and may allow some quantitation (49), but there may be problems in gaining access to the basolateral surface, and suitable controls must be used.

ii. Post-embedding methods

Frozen ultrathin sections

This is a very sensitive method, combined with potentially good morphology (50), but specialized, expensive equipment and a high degree of technical skill are necessary (51).

Hydrophilic plastic resin

The cells are embedded, for example in Lowicryl resin (52). The main problem with this technique is that because of the limited accessibility to antigen, labelling for membrane proteins may be scarce. Recently, this method has been combined with freeze-substitution with some success (53, 54).

NB For further details on immunocytochemistry for the electron microscope, the reader is referred to Griffiths (55).

5. Biochemical methods

5.1 Polarized internalization of transferrin

Transferrin mediates the cellular uptake of iron from serum by receptor-mediated endocytosis. In the low pH of the endosomal compartment, the iron dissociates and the apotransferrin recycles back to the plasma cell surface where it leaves its receptor (56). Most epithelial cells express the transferrin receptor on the basolateral surface. An exception is the human placental trophoblast, as illustrated by the BeWo cell line (57), which expresses the receptor on both surfaces. A method for measuring the polarized uptake of transferrin is detailed in *Protocol 7* and sample results are displayed in *Figure 6*.

Protocol 7. To test for polarized uptake of [^{125}I]transferrin

Equipment and reagents

- 8 × 24 mm Transwells with MDCK cells
- Unlabelled canine transferrin (TF) (Sigma)
- [^{125}I]TF; can be labelled with the chloramine-T method (58)
- Serum-free (SF) medium; prewarmed DMEM-0.2% BSA (w/v)
- Ice-cold PBSB: PBSA containing 0.1 mM $CaCl_2$, 1 mM $MgCl_2$ and 0.2% (w/v) BSA
- Gamma counter
- Disposable scalpel
- LP3 tubes (Marathon Laboratory Supplies)

Method

1. Wash the cells twice with SF-medium and re-incubate for 1 h to 'chase out' the intracellular TF.
2. Remove the medium and set up duplicate filters as follows (A, apical chamber; B, basolateral chamber):
   (a) A: 1.5 ml SF-medium, B: 1.5 ml SF-medium plus 5×10^5 c.p.m. [^{125}I]TF
   (b) A: 1.5 ml SF-medium, B: 1.5 ml SF-medium plus 5×10^5 c.p.m. [^{125}I]TF plus 100 μg/ml unlabelled TF
   (c) A: 1.5 ml SF-medium plus 5×10^5 c.p.m. 1 [^{125}I]TF, B: 1.5 ml SF-medium
   (d) A: 1.5 ml SF-medium plus 5×10^5 c.p.m. [^{125}I]TF plus 100 μg/ml unlabelled TF, B: 1.5 ml SF-medium.
3. Place the cells in an appropriate container (leaded Pyrex) and replace in the incubator for 90 min. This is long enough to allow intracellular TF receptors to become labelled.
4. Place the cells on ice, and wash them about eight times with cold PBSB to remove non-specifically bound TF.
5. Cut out the filters, place in LP3 tubes, and count them in a gamma counter.
6. Subtract values (b) from (a) and (d) from (c) to give the specific uptake at the basolateral and apical surfaces, respectively.

5.2 Selective cell surface labelling of membrane proteins

Cell surface proteins from epithelial cells grown as polarized monolayers on permeable supports can be selectively labelled on either the apical or the basolateral membrane (59). A number of protocols have been used to differentially tag apical or basolateral proteins (31); cell surface biotinylation is presented in *Protocol 8*, and cell surface radio-iodination in *Protocol 9*.

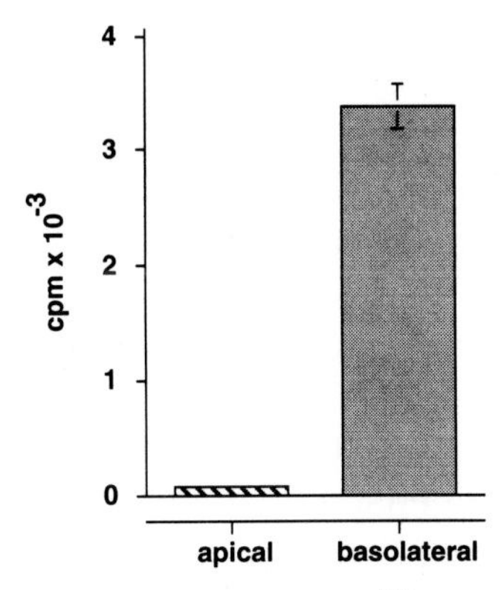

Figure 6. Graph showing the polarized uptake of [^{125}I]transferrin by MDCK I cells. In MDCK cells, the transferrin receptor is almost entirely located on the basolateral membrane (23).

5.2.1 Cell surface biotinylation

MDCK cells grown on filters can be biotinylated as described (60, 61). This is currently the most widely used method to selectively tag membrane proteins from either domain. The major advantage of the method is that biotinylated protein can be detected (e.g. with streptavidin–HRP by Western blot) and/or isolated (with streptavidin–agarose) (31, 61).

Protocol 8. Cell surface biotinylation on filter-grown cells

Equipment and reagents

- Filter-grown cells (see *Protocol 2*)
- PBSC/M (pH 7.4): mix 150 mM NaCl, 10 mM KH_2PO_4, 4 mM KCl, 0.1 mM $CaCl_2$, and 1 mM $MgCl_2$; adjust to pH 7.4. Pass the solution through a 0.22 μm filter and store at 4°C
- PBSC/M (pH 8.4) as above, with pH adjusted to 8.4
- Sulfo-NHS-biotin (Pierce): dissolve in DMSO at 200 mg/ml and store at −20°C in small aliquots (e.g. 20 μl)
- Protein-free DMEM
- Biotinylation solution (to be prepared just before cell surface biotinylation): dissolve sulfo-NHS-biotin at 0.5 mg/ml (1/400 dilution from stock) in ice-cold PBSC/M (pH 8.4). The addition of sulfo-NHS-biotin–DMSO will reduce the pH to about 8
- A shaker with variable shaking speed placed in a cold room (4 °C)
- Lysis buffer (for immunoprecipitation; IPP): mix 60 mM EDTA, 10 mM Tris–HCl (pH 7.4), 0.4% deoxycholate (DOC), and 1% Nonidet P-40. Just before use, add a cocktail of protease inhibitors (1/1000 dilution) (17.4 mg/ml PMSF in ethanol, 5 mg/ml of leupeptin, pepstatin, and antipain in DMF, 10 mg/ml soybean trypsin inhibitor in water; all from Sigma)
- TENT-1%: mix 50 mM Tris–HCl (pH 7.5), 5 mM EDTA, 150 mM NaCl, and 1% Triton X-100
- Primary or secondary antibody linked to Sepharose beads (see ref. 62)
- Streptavidin–agarose (Pierce)
- Protein sample buffer (SB 2 ×): mix 160 mM Tris–HCl (pH 6.9), 25% (w/v) sucrose, 7% (w/v) SDS and 0.012% (w/v) Bromophenol Blue
- Bench centrifuge (at 4°C)

Protocol 8. *Continued*

Method

NB All steps are performed on ice with ice-cold solutions; working in a cold room (4°C) is an advantage. Remember that basolateral medium is removed first and apical medium is added first to avoid hydrostatic pressure from the basal chamber.

1. After the determined culture time (see *Protocol 2*), transfer the filters to ice.
2. Rinse both filter compartments three times with ice-cold PBSC/M (pH 7.4).
3. Add 1 ml and 1.5 ml PBSC/M (pH 7.4) to the apical and basal compartments, respectively, and incubate with shaking for 30 min.
4. Remove PBSC/M. To the domain to be labelled, add 1 ml (apical) or 1.5 ml (basolateral) of biotinylation solution and add the corresponding volume of PBSC/M (pH 7.4) to the unlabelled domain. Incubate with shaking for 30 min in a cold room (set the shaking speed so that the filter device is stable on the shaker).
5. Wash the filters once with PBSC/M (pH 7.4), followed by one wash with DMEM (protein-free) and three washes with PBSC/M (pH 7.4) in order to remove and quench all reactive sulfo–NHS–biotin groups.
6. Prepare the appropriate cell lysate for either total protein analysis or IPP/streptavidin–agarose purification of biotinylated proteins.
7. For IPP and streptavidin–agarose purification, lyse the cells with 500 μl of lysis buffer and use a rubber policeman to scrape off the cells.[a]
8. Clear the lysate by centrifugation at 10 000 *g* for 5 min.[b] Transfer the supernatant into a 5 ml tube with a tight cap and add 4.5 ml of TENT-1%. Add the antibody–Sepharose or the streptavidin–agarose and incubate at 4°C overnight or for 2 h (at least) for the respective purification step.
9. Spin the beads for 1 min at 1000 *g* and transfer beads with 1 ml TENT-1% into a 1.5 ml Eppendorf tube. Wash beads four times with 1 ml TENT-1% and elute the proteins from the antibody with 100 μl SB by heating at 95°C for 5 min.[c]
10. Spin the beads for 5 min at 10 000 *g* and collect the supernatant. Load on SDS–PAGE in either non-reducing or reducing conditions.

[a] Alternatively, the filter can be cut out with a razor blade and the cells lysed in a well with the lysis buffer, with shaking (60). This avoids the lysis of the cells grown on the plastic support which are not as polarized as those in direct contact with the filter (*Figure 2*) (63).

[b] At this stage, the cell lysate can be incubated at 95 °C for 5 min in the presence or absence of SDS (3% final concentration) in order to destroy protease activities and denature the antigen to facilitate antibody accessibility to the epitope. Then continue as described.

[c] To elute biotinylated proteins from the streptavidin–agarose beads, boil the sample three times, for 5 min each time, in the presence of 10 mM DTT. Resolve protein on SDS–PAGE and detect protein with a suitable method (see *Figure 7* for some examples).

Following cell surface biotinylation, cell lysates are prepared for either total protein or protein IPP analysis as illustrated in *Figure 7A* (60, 61). Alternatively, cells can be incubated with completed medium at 37 °C in the CO_2 incubator for chase experiments, where the fate of biotinylated cell surface proteins can be followed (endocytosis, transcytosis, or secretion). Biotinylated proteins can be detected by Western blot (60, 61) (for general Western blot protocols, see Harlow and Lane, ref. 64). To follow the polarized cell surface appearance of newly synthesized proteins (pulse–chase experiments), biosynthetically labelled cells grown on filters are biotinylated following pulse labelling and chase, and the protein of interest is isolated by serial purification steps as shown in *Figure 7B* (65, 66).

IPP proteins are eluted from antibodies, diluted in buffer (*Protocol 8*), and biotinylated proteins (from either the apical or basolateral cell surface) are recovered with streptavidin–agarose. Non-biotinylated proteins (intracellular pool or from the opposite membrane) can be re-immunoprecipitated from the supernatant with the same antibody used in the primary purification step in order to quantify the proportion of the different cellular domain-associated pools (*Figure 7*). A combination of different cell surface labelling methods (biotinylation and exogalactosylation) has also been used to follow endogenous protein transcytosis in MDCK cells (36).

Cell surface radio-iodination does have its limitations, such as dependence of the presence of tyrosine residues on the protein to be labelled and accessibility of the basolateral domain to enzymes (lactoperoxidase and glucose oxidase, see *Protocol 9*) (e.g. ref. 60); however, the biotinylation method has been shown to have disadvantages of a similar nature (63). Gottardi and Caplan (63) present experimental evidence for such limitations of biotinylation and they discuss the importance of assessing the limitations of both the methods and the model system (including the protein of interest and the cell line used) one is working with. For additional comparison of additional methods to assess cell surface polarity, see ref. 61.

5.2.2 Cell surface radio-iodination

Cell surface radio-iodination was the first method to be used to investigate the steady-state distribution of proteins on epithelial cells grown on filters (59). Since this method can give different results from those obtained with cell surface biotinylation, we describe the method in *Protocol 9*. We would suggest that they can be used in parallel to obtain complementary results. In MDCK cells, apical membrane proteins seem to be efficiently labelled with the two methods, since both protocols produced similar results when the distribution of glycosyl-phosphatidylinositol (GPI)-anchored proteins was investigated (67, 68).

Figure 8 shows typical results for two proteins with different cellular distribution labelled by radio-iodination. The labelling was performed on transfected MDCK I cells expressing either the pIg-R (found on both domains) or

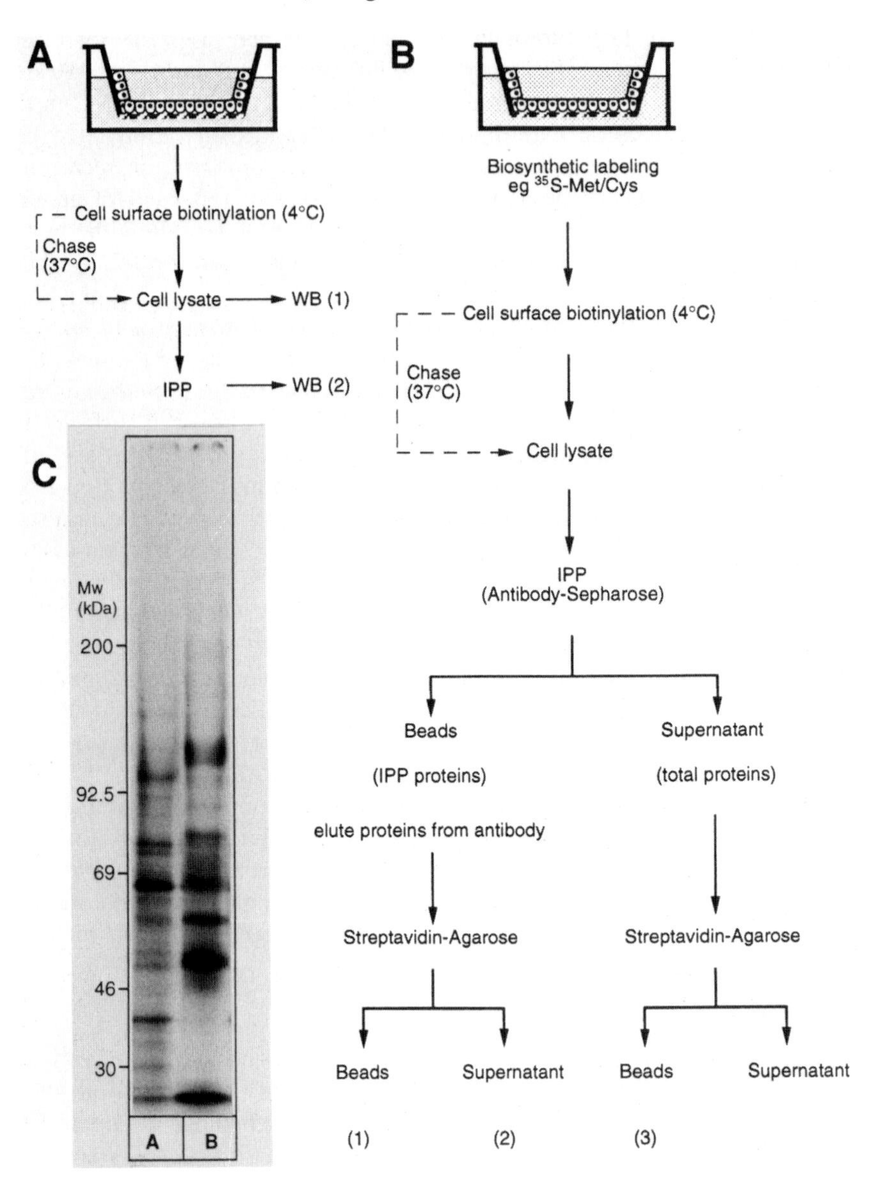

Thy-1 (apical) antigens (65, 68). In this system the expression is dependent on the stimulation of a glucocorticoid analogue, dexamethasone (DEX) (32). As shown in *Figure 8A*, the presence of the pIg-R antigen on both surfaces is dependent on stimulation with DEX. The results also suggest that the band observed at about 200 kDa is specific. This 200 kDa polypeptide could be either a homodimer of the pIg-R (69) or a heterodimer between the pIg-R and an unknown protein, specific for the basolateral membrane. For further

Figure 7. Use of cell surface biotinylation to study the distribution and trafficking of membrane proteins in filter-grown cells. (A) General scheme to identify membrane proteins associated with either domain without using radioisotopes. The biotinylated proteins can be detected by Western blot (WB) using streptavidin–HRP with chemiluminesence. Steady-state distribution of total protein can be detected (sample 1) or alternatively, a specific protein can be immunoprecipitated (IPP) and the cell surface pool subsequently detected (sample 2). A chase can be performed post-biotinylation. This allows, for example, the fate of a protein trafficking between different domains to be followed. (B) Study of the fate of newly synthesized membrane proteins which are expressed on the cell surface and/or involved in endocytosis and transcytosis. Biosynthetically labelled cells can be biotinylated and the radiolabelled proteins identified by serial purification steps. This allows different cellular pools of a given protein to be isolated (1, cell surface pool of the IPP protein; 2, pool of non-biotinylated IPP protein, i.e. intracellular or from the non-labelled domain). In the same experiment total cell surface protein can also be recovered (sample 3) and compared for the two domains, and used as an internal control for the specificity of the labelling (tightness-dependent) and the polarized state of the cell. (C) Represents a typical pattern of total proteins labelled for 15 min with [^{35}S]cysteine and chased for 2 h corresponding to sample 3 in (B). Following elution from the beads, the proteins were resolved on a 5–13% SDS–PAGE under reducing conditions. A distinct pattern for the two domains can be seen indicating that the cells are polarized (A and B for apical and basolateral biotinylation, respectively).

---

details on the expression system, transfection method, and the two cell lines expressing the heterologous proteins see refs 32, 69, and 70.

Protocol 9. Cell surface radio-iodination of filter-grown cells

NB A knowledge of working with radioisotopes is assumed

Equipment and reagents

- Filter-grown cells (see *Protocol 2*)
- PBS–glucose/calcium (PBSG/C): mix 85 mM NaCl, 17.5 mM $NaHCO_3$, 4 mM KCl, 0.8 mM KH_2PO_4, 1 mM $CaCl_2$ and 10 mM glucose. Prepare the solution 24 h in advance to allow the L-glucose to become D-glucose which is the substrate for glucose oxidase.
- Lactoperoxidase (Sigma): 10 mg/ml in PBS and stored at – 20 °C in 100 μl aliquots
- Glucose oxidase (GO) (Sigma type 5): stored at 4°C and diluted 1:100 in water just before use
- Labelling mix: PBSG/C containing 1 mg/ml lactoperoxidase and 1 mCi carrier-free [^{125}I] NaI (100 mCi/ml; Amersham, UK)
- 20% sodium azide (freshly prepared)
- PBSG/C containing 0.2% sodium azide (freshly prepared)

Method

1. Transfer the filter to ice and rinse three times with ice-cold PBSG/C.
2. Add 1 ml of ice-cold labelling mix to the domain to be labelled and an equal volume of PBSG/C to the opposite domain. Incubate on ice for 30 min to allow diffusion of enzymes and [^{125}I]NaI.
3. Start the labelling reaction by adding 20 μl of diluted GO to the side to be radio-iodinated. Incubate on ice for 10–30 min.

Protocol 9. Continued

4. Stop the reaction by adding 20 μl of 20% sodium azide to both sides. Count 2.5 μl from the labelled domain and 50 μl from the opposite side to determine the integrity of the filter during the labelling procedure.[a]
5. Rinse the monolayers six times with ice-cold PBSG/C–0.2% sodium azide and process the filter for IPP. A 10 μl aliquot of the total cell lysate can be precipitated with TCA in order to assess the total pattern of radio-iodinated proteins (68). This gives additional information on the specificity of the labelling and the polarity of the cells.

[a]Usually ratios of 1:100 to 1:200 are obtained with MDCK I cells (R. Hirt, personal observation).

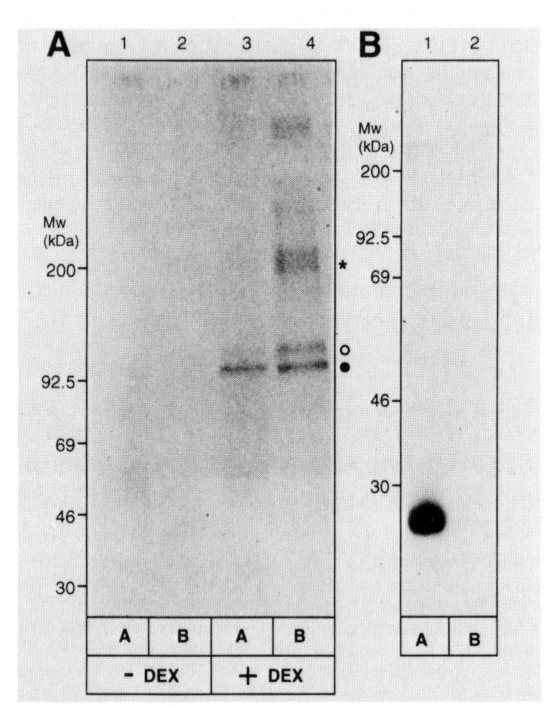

Figure 8. Use of radio-iodination to analyse cell surface polarity. Transfected MDCK cells expressing the pIg-R and the Thy-1 antigen were grown on filters for three days and stimulated for 16 h with 1 μM DEX. Cells were labelled either from the apical (A) or basolateral (B) side as indicated in *Protocol 9.* (A) Immunoprecipitation of the pIg-R using an anti-tail monoclonal antibody (71) from non-induced (lanes 1 and 2) and DEX-induced cells (lanes 3 and 4). The position of the different M_w forms of the pIg-R differentially present in the two domains are indicated by a closed circle (apical and basolateral), an open circle, and an asterisk (basolateral) (5–13% SDS–PAGE). (B) as in (A), but with a cell line expressing Thy-1 and using a rabbit polyclonal anti-Thy-1 antibody (gift from Prof. C. Bron, Institute of Biochemistry, University of Lausanne, Switzerland) for IPP (10% SDS–PAGE). For more details on transfected MDCK cell lines expressing the pIg-R and Thy-1 antigens see refs 32 and 70.

6. Conclusions

Since protocols to grow epithelial cells on permeable support were described (15), numerous morphological and biochemical approaches have been used to study the steady-state distribution and dynamic trafficking of proteins in polarized epithelial cells. When used in combination with molecular biology techniques allowing the expression of wild type and mutant heterologous proteins in epithelial cell lines (72), these methods have allowed the characterization of some of the sorting signals which direct the membrane proteins around epithelial cells (12). Recent work suggests that data accumulated for epithelial cells may have widespread implications for our understanding of membrane trafficking in non-epithelial cells (12) such as neurons (13, 73). In addition, results obtained from the genetic model *Drosophila* have allowed the identification of new key players (homologous between vertebrate and insects) in the establishment and maintenance of epithelial cell polarity (74). In this chapter we have presented a sample of techniques available to study cell surface polarity. The reader is encouraged to follow up the references to obtain further information in order to select the appropriate method(s) to answer their particular question(s) of interest.

Acknowledgements

The authors are grateful to Professors C. R. Hopkins (MRC Laboratory for Molecular Cell Biology, London, UK) and J.-P. Kraehenbuhl (Institute of Biochemistry and ISREC, Lausanne, Switzerland) for support in the experimental work carried out in their laboratories. This work was supported by the Swiss National Science Foundation and a British Council/Swiss Government travel grant to C.R.H. and J.-P.K. We would also like to thank M. Shipman and L. Racine for their excellent technical assistance, Drs C. Isacke, H. Sheikh, and Robert Parton for making available their micrographs, and C. Connolly and M. Embley for critically reading the manuscript.

References

1. Alberts, B., Bray, D., Lewis, J., Raff, M., Roberts, K., and Watson, J. D. (1994). *Molecular biology of the cell* (3rd edn). Garland Publisher Inc., New York and London.
2. Simons, K. and Fuller, S. D. (1985). *Annu. Rev. Cell Biol.*, **1**, 243.
3. Rodriguez-Boulan, E. and Nelson, W. J. (1989). *Science*, **245**, 718.
4. Kraehenbuhl, J. P. and Neutra, M. R. (1992). *Physiol. Rev.*, **72**, 853.
5. Edelman, G. M. (1983). *Science*, **219**, 450.
6. Straehelin, L. A. (1974). *Int. Rev. Cytol.*, **39**, 191.
7. Anderson, J. M., Balda, M. S., and Fanning, A. (1993). *Curr. Opin. Cell Biol.*, **5**, 772.

8. Schwarz, M. A., Owaribe, K., Kartenbeck, J., and Franke, W. W. (1990). *Annu. Rev. Cell Biol.*, **6**, 461.
9. Loewenstein, W. R. (1986). *Cell*, **48**, 726.
10. Fey, E. G., Wan, K., and Pennans, S. (1984). *J. Cell Biol.*, **98**, 1973.
11. Mays, R. W., Beck, K. A., and Nelson, W. J. (1994). *Curr. Opin. Cell Biol.*, **6**, 16.
12. Matter, K. and Mellman, I. (1994). *Curr. Opin. Cell Biol.*, **6**, 545.
13. Rodriguez-Boulan, E. and Powell, S. K. (1992). *Annu. Rev. Cell Biol.*, **8**, 395.
14. Mostov, K., Apodaca, G., Aroeti, B., and Okamoto, C. (1992). *J. Cell Biol.*, **116**, 577.
15. Cereijido, M., Robbins, E. S., Dolan, W. J., Rotunno, C. A., and Sabatini, D. D. (1978). *J. Cell Biol.*, **77**, 853.
16. Fuller, S., von Bonsdorff, C. -H., and Simons, K. (1984). *Cell*, **38**, 65.
17. ATCC (1981). *American Type Culture Collection: Catalogue of Strains* II, 52. ATCC, Rockville, MD.
18. Richardson, J. C. W., Scalera, V., and Simmons, N. L. (1981). *Biochim. Biophys. Acta*, **673**, 26.
19. Rodriguez-Boulan, E. and Pendergast, M. (1980). *Cell*, **20**, 45.
20. Balcarova-Ständer, J., Pfeiffer, S. E., and Simons, K. (1984). *EMBO J.*, **3**, 2687.
21. Rindler, M. J., Chuman, L., Shaffer, L., and Sajier, M. H. (1979). *J. Cell Biol.*, **81**, 635.
22. Stevenson, B. R., Anderson, J. M., Goodenough, D. A., and Mooseker, M. S. (1988). *J. Cell Biol.*, **107**, 2401.
23. Fuller, S. D. and Simons, K. (1986). *J. Cell Biol.*, **103**, 1767.
24. Hansson, G. C., Simons, K., and van Meer, G. (1986). *EMBO J.*, **5**, 483.
25. Mostov, K. and Deitcher, D. L. (1986). *Cell*, **46**, 613.
26. Freshney, R. I. (ed.) (1986). *Animal cell culture: a practical approach*. IRL Press, Oxford.
27. Matlin, K. S. and Simons, K. (1983). *Cell*, **34**, 233.
28. Caplan, M. J., Stow, J. L., Newman, A. P., Madri, J., Anderson, H. C., Farquhar, M. G., Palade, G. E., and Jamieson, J. D. (1987). *Nature*, **329**, 632.
29. Patrone, L. M., Cook, J. R., Crute, B. E., and Buskirk, R. G. (1992). *J. Tiss. Cult. Meth.*, **14**, 225.
30. Hay, E. D. (1993). *Curr. Opin. Cell Biol.*, **5**, 1029.
31. Ellis, J. A., Jackman, M. R., Perez, J. H., Mullock, B., and Luzio, J. P. (1992). In *Protein targeting: a practical approach* (ed. A. I. Magee and T. Wileman), pp. 25–57. IRL Press, Oxford.
32. Hirt, R. P., Poulain-Godefroy, O., Billotte, J., Kraehenbuhl, J. P., and Fasel, N. (1992). *Gene*, **111**, 199.
33. Schoenenberger, C. A., Zuk, A., Kendal, D., and Matlin, K. S. (1991). *J. Cell Biol.*, **112**, 873.
34. Kondor-Koch, C., Bravo, R., Fuller, S. D., Cutler, D., and Garoff, H. (1985). *Cell*, **43**, 297.
35. Vega-Salas, D. E., Salas, P. J. L., Gundersen, D., and Rodriguez-Boulan, E. (1987). *J. Cell Biol.*, **104**, 905.
36. Brändli, A. W., Parton, R. G., and Simon, K. (1990). *J. Cell Biol.*, **111**, 2902.
37. Shore, E. M. and Nelson, W. J. (1991). *J. Biol. Chem.*, **266**, 19672.
38. Penn, E. J., Hobson, C., Rees, D. A., and Magee, A. I. (1987). *J. Cell Biol.*, **105**, 57.

39 Hughson, E. J., Cutler, D. F., and Hopkins, C. R. (1989). *J. Cell Sci.*, **94**, 327.
40. Hughson, E. J. and Hopkins, C. R. (1990). *J. Cell Biol.*, **110**, 337.
41. Louvard, D. (1980). *Proc. Natl Acad. Sci. USA*, **77**, 4132.
42. Matsumoto, B. (ed.) (1993). *Methods in cell biology*, Vol. 38. Academic Press Inc., San Diego.
43. Kendall, D., Lencer, W., and Matlin, K. S. (1992). *J. Tiss. Cult. Meth.*, **14**, 181.
44. Asai, D. J. (ed.) (1993). *Methods in cell biology*, Vol. 37. Academic Press, Inc., San Diego.
45. Reynolds, L. S. (1963). *J. Cell Biol.*, **17**, 208.
46. Glauert, A. M. (ed.) (1975). *Practical methods in electron microscopy*. Vol. 3. North Holland Publishing Company, Amsterdam.
47. Dykstra, M. J. (ed.) (1993). *A manual of applied techniques for biological electron microscopy*. Plenum Press, London and New York.
48. Matlin, K., Bainton, D. F., Pesonen, M., Louvard, D., Genty, N., and Simons, K. (1983). *J. Cell Biol.*, **97**, 627.
49. Wessels, H. P., Hansen, G. H., Fuhrer, C., Look, A. T., Sjostrom, H., Noren, O., and Spiess, M. (1990). *J. Cell Biol.*, **111**, 2923.
50. Klumperman, J., Fransen, J. A. M., Tager, J. M., and Ginsel, L. A. (1992). *Eur. J. Cell Biol.*, **57**, 147.
51. Tokuyasu, K. T. (1980). *Histochem. J.*, **12**, 381.
52. Roth, J., Bendayan, M., Carleman, E., Villiger, W., and Garavito, M. (1981). *J. Histochem. Cytochem.*, **29**, 663.
53. Harris, D. S., Slot, J. W., Geuze, H. J., and James, D. E. (1992). *Proc. Natl Acad. Sci. USA*, **89**, 7556.
54. van Genderen, I. L., van Meer, G., Slot, J. W., Geuze, H. J., and Voorhout, W. F. (1991). *J. Cell Biol.*, **115**, 1009.
55. Griffiths, G. (1993). *Fine structure immunocytochemistry*. Springer–Verlag, New York.
56. Klausner, R. D., Ashwell, G., van Renswoude, J., Harford, J. B., and Bridges, K. R. (1983). *Proc. Natl Acad. Sci. USA*, **80**, 2263.
57. Cerneus, D. P. and van der Ende, A. (1991). *J. Cell Biol.*, **114**, 1149.
58. Hunter, W. M. and Greenwood, F. C. (1962). *Nature*, **194**, 495.
59. Richardson, J. C. W. and Simmons, N. L. (1979). *FEBS Lett.*, **105**, 201.
60. Sargicomo, M., Lisanti, M. P., Graeve, L., LeBivic, A., and Rodriguez-Boulan, Boulan, E. (1989). *J. Membrane Biol.*, **107**, 277.
61. Rodriguez-Boulan, E., Salas, P. J., Sargicomo, M., Lisanti, M., LeBivic, A., Sambuy, U., Vegas-Salas, D., and Graeve, L. (1990). In *Methods in cell biology* (ed. A. M Tartakoff), Vol. 32, pp. 37–56. Academic Press, NY.
62. Otto, J. J. and Seung-won, L. (1993). *Methods in cell biology* (ed. B. Matsumoto), Vol. 38, pp. 119–29. Academic Press Inc., San Diego.
63. Gottardi, C. J. and Caplan, M. J. (1992). *J. Tiss. Cult. Meth.*, **14**, 173.
64. Harlow, E. and Lane, D. (1988). *Antibodies, a laboratory manual.* Cold Spring Harbor Laboratory Press, NY.
65. Le Bivic, A., Sambuy, Y., Mostov, K., and Rodriguez-Boulan, E. (1990). *J. Cell Biol.*, **110**, 1533.
66. Matter, K., Brauchbar, M., Bucher, K., and Hauri, H.-P. (1990). *Cell*, **60**, 429.
67. Lisanti, M. P., Sargiacomo, M., Graeve, L., Saltiel, A. R., and Rodriguez-Boulan, E. (1988). *Proc. Natl Acad. Sci. USA.*, **85**, 9557.

68. Wilson, J. M., Fasel, N., and Kraehenbuhl, J. -P. (1990). *J. Cell Sci.*, **96**, 143.
69. Hirt, R. P., Hughes, G. J., Frutiger, S., Michetti, P., Perregaux, C., Poulain-Godefroy, O., Jeanguenat, N., Neutra, M. R., and Kraehenbuhl, J. P. (1993). *Cell*, **74**, 245.
70. Hirt, R. P., Fasel, N., and Kraehenbuhl, J. P. (1994). In *Methods in cell biology* (ed. M. G. Roth), Vol. 43, pp. 247–62. Academic Press, San Diego, CA.
71. Solari, R., Kühn, L., and Kraehenbuhl, J. P. (1985). *J. Biol. Chem.*, **260**, 1141.
72. Roth, M. G. (1989). In *Methods in cell biology* (ed. K. S. Matlin and J. D. Valentich). Vol. 8, pp. 269–302. A. R. Liss, NY.
73. Simons, K., Dupree, P., Fiedler, K., Huber, L. A., Kobayashi, T., Kurzchalia, T., Olkonnonen, V., Pimplikar, S., Parton, R., and Doti, C. (1992). *Cold Spring Harbor Symp. Quant. Biol.,* Vol. 57, 611.
74. Knust, E. (1994). *Trends Genet.*, **10**, 275.

4

Xenobiotic metabolism in epithelial cell cultures

ANDRÉ GUILLOUZO and CHRISTOPHE CHESNÉ

1. Introduction

Individuals are exposed to a variety of xenobiotics which are present in their environment (i.e. air, water, etc.), are administered (i.e. medicines), or are ingested (i.e. beverages and foods). Many of these compounds are lipophilic and must be converted to more hydrophilic metabolites in order to be eliminated. Most, if not all epithelial cells, contain some capacity to metabolize foreign compounds, however hepatocytes contain the most diverse and active xenobiotic metabolizing system. This explains why the hepatic enzyme profile has been the most extensively analysed and why it is critical to determine hepatic biotransformation of new compounds that could reach the systemic circulation. To gain this information, *in vitro* liver models and particularly cultured liver cells are increasingly used.

A number of substrates can be used to estimate the amount of the different enzymes involved in either phase I and phase II reactions. Depending on the enzyme, they can be measured on living cells, homogenates, and/or cytosolic or microsomal fractions. In this chapter we outline the basic conditions for culturing hepatocytes, and provide detailed protocols for determining the activities of various phase I and phase II metabolizing enzymes. These protocols can also be used for the determination of drug metabolizing enzyme activities in hepatic cell lines and in non-hepatic cells, e.g. keratinocytes, however, depending on the enzyme concentration, it may be necessary to modify the protocols usually by using higher cell numbers. The protocols described correspond to the most frequently tested enzyme activities and cover the main experimental approaches used, i.e. spectrophotometry, fluorimetry, HPLC, and TLC.

2. Hepatocyte cultures

2.1 Primary hepatocyte cultures

The two-step collagenase perfusion method is now routinely used to prepare isolated hepatocytes from either whole organs, lobes, or wedge biopsies.

Parenchymal cells can be obtained from the liver of humans and various animal species. Detailed protocols are described in several reviews or books (1–3). Isolated hepatocytes can be used immediately in suspension but they do not survive for more than a few hours. Different basal media, e.g. RPMI 1640, BME, Ham's F-12, and Williams' E, are commonly used for short-term culturing of hepatocytes with only small differences in cell survival and albumin secretion rate being observed (3). To survive longer, hepatocytes must be cultured. The conditions used for culturing hepatocytes (e.g. culture medium, seeding cell density, substrate, etc.) vary from one laboratory to another and sophisticated conditions are required for long-term compared to short-term cultures (3). Hepatocyte survival and function can be improved by:

(a) the addition of various soluble factors to the culture medium;
(b) the use of matrix proteins which allow the cells to retain a globular shape, e.g. Matrigel (see Chapter 7, *Protocol 2*);
(c) culturing the cells as aggregates; and
(d) co-culturing the cells with non-parenchymal cells, e.g. rat liver epithelial cells (RLEC) of biliary origin (see *Protocol 1*). This method was developed by Guguen-Guillouzo *et al.* (4) and appears to be the most powerful system for *in vitro* long-term maintenance of liver-specific functions (5).

These different culture conditions are described in detail in several reviews (2, 3, 6).

Protocol 1. Co-culture of hepatocytes and RLEC

Equipment and reagents

- 10 livers from 8–10 day old rats
- Hepes buffer: 160.8 mM NaCl, 3.15 mM KCl, 0.7 mM $NaPO_4 \cdot 12H_2O$, and 33 mM Hepes in distilled water (pH 7.6)
- 0.25% trypsin solution: 0.25% (w/v) trypsin and 5% (v/v) FCS in Hepes buffer
- Williams' E medium (pH 7.4) (Gibco)
- Hepatocytes[a]
- 0.075% collagenase solution: 0.075% (w/v) collagenase in Hepes buffer (pH 7.6)
- 35 mm plastic culture dishes (Nunc)
- Culture medium A: 3:1 mixture of Minimum Essential Medium (MEM; Gibco) and medium 199 (Gibco) supplemented with 1 mg/ml BSA (Sigma), 5 μg/ml bovine insulin (Sigma), 50 μg/ml streptomycin, 50 IU/ml penicillin, and 10% FCS (pH 7.4) (J. BIO).
- Culture medium B: medium A with the addition of 3.5 μM hydrocortisone hemisuccinate (pH 7.4)
- 5% CO_2/95% air humidified incubator

A. *Isolation of RLEC according to Williams* et al. *(7)*

1. Cut the livers into small fragments and incubate for 15 min at 37 °C in 50 ml 0.25% trypsin solution. Carefully pour off the solution and replace with fresh trypsin solution. Incubate at 37 °C for a further 15 min.
2. Gently disaggregate the fragments and centrifuge the cell suspension at 50 *g* for 2 min. Wash the cells twice by resuspending the cells in

20 ml Hepes buffer and centrifuging at 50 *g* for 2 min. Resuspend the cells in 20 ml Williams' E medium.

3. Seed the cells into a cell culture dish. After 20 min, collect the supernatant (containing the unattached cells) and seed into a new dish. Repeat this step three times so that the third and fourth dishes contain mainly epithelial cells.[b,c]

B. *Co-culture of hepatocytes and RLEC*

1. Plate 8×10^5 hepatocytes in 2 ml of culture medium A in a 35 mm diameter culture dish.[c,d] Incubate the cells at 37 °C under a humidified atmosphere of 5% CO_2/95% air.
2. Once the cells have attached,[e] remove the medium (including any dead cells) and add 1×10^6 RLECs in fresh medium. Reincubate the cells for 24 h after which they should have reached confluence.[f]
3. Renew the medium after 24 h and every day thereafter with culture medium B[g].

[a] Freshly isolated hepatocytes can be separated into subpopulations according to their degree of ploidy or their intralobular distribution. Before seeding, hepatocytes can be either hypothermically preserved for a few days or cryopreserved for weeks or even years (see refs 3 and 6).
[b] Contaminating fibroblasts are eliminated as they attach to the cell culture plastic more rapidly than the epithelial cells. The remaining contaminating cells can be eliminated by selective killing with a sterile needle or by epithelial cell cloning.
[c] RLECs can be maintained either by subculturing in Williams' E medium containing 5–10% FCS, or by cryopreservation.
[d] The dishes may or may not be coated with collagen.
[e] After approximately 3 h, animal liver cells will have attached to the plastic and begun to spread. A longer period is usually required for the attachment of human hepatocytes.
[f] Alternatively, hepatocytes may be seeded on to a confluent monolayer of RLECs.
[g] Hepatocytes can be selectively detached from RLECs by incubating in 0.075% collagenase solution for 10 min (3).

2.2 Hepatic cell lines

Hepatic cell lines can be obtained from hepatoma cells and from normal hepatocytes which have been immortalized by viral or cellular oncogenes. None of the hepatic cell lines obtained up to now express the majority of phase I and phase II drug reactions. Only a few enzymes are present, if any, and their levels are usually lower than those measured in their normal counterparts.

3. Preparation of samples

A number of methods, first designed for the measurement of drug metabolizing enzyme activities in microsomes, have been adapted for intact cells. Various enzymes can currently be measured directly by incubating living cell cultures with specific substrates. In contrast to microsomal preparations (see *Protocol 2*), which are difficult to prepare from non-hepatic cells, e.g. keratinocytes, a relatively small number of cells are frequently sufficient for these assays. For example, 3×10^4 liver parenchymal cells can be used for the determination of ethoxyresorufin *o*-deethylase activity. Furthermore, these techniques are often easier to perform with whole cells than with cell homogenates or microsomes.

Protocol 2. Preparation of microsomes from cultured hepatocytes

Equipment and reagents

- Cultured hepatocytes (approximately 2×10^7 cells)
- Homogenizing medium: 100 mM (12.11 g/litre) Tris buffer containing 250 mM saccharose and 1 mM EDTA. Adjust to pH 7.4 using HCl
- Homogenizer
- Ultracentrifuge
- Freezing medium: 20% (v/v) glycerol in 0.1 M phosphate buffer (pH 7.4)
- PBS (Gibco)

Method

1. Remove the culture medium from the cells and wash them once with PBS.
2. Scrape the cells off the plastic and collect them in 5 ml of homogenizing medium.[a]
3. Homogenize the cell suspension for 30 sec.
4. Centrifuge the homogenate at 800 *g* for 10 min. Remove the supernatant and centrifuge it again at 800 *g* for 10 min.
5. Remove the supernatant and centrifuge it at 13 000 *g* for 20 min. Remove the 'post-mitochondrial' supernatant and centrifuge it at 105 000 *g* for 60 min.
6. Resuspend the microsomal pellet in freezing medium (10–30 mg protein/ml) and store at –80 °C until required.

[a] For the remainder of the protocol, the biological material should be kept at 4 °C.

4. Determination of drug metabolizing capacity

Total cytochrome P450 (CYP) content and various drug metabolizing enzyme activities can be measured in order to estimate the xenobiotic metabolic capacity of cells in culture. The methods used for measuring total CYP (see *Protocol 3*) and the most frequently selected enzyme activities are described here. These enzymes represent phase I reactions, catalysed by the most important CYP isozymes (see *Protocols 4–10*), or phase II reactions (see *Protocols 11–14*).

Protocol 3. Determination of total CYP content[a]

Equipment and reagents

- Cell monolayers in 60 or 90 mm diameter culture dishes (approximately 10^5 cells/cm^2)
- Emulgen buffer (pH 7.4): 0.1 M NaH_2PO_4 (Merck), 0.1 M Na_2HPO_4 (Merck), 1 mM EDTA, 0.1% (w/v) Emulgen 913 (Kao corporation, Japan), and 20% (v/v) glycerol
- PBS
- Narrow-bore glass pipette
- Sodium dithionite (Fluka)
- Carbon monoxide (situated by an efficient fume hood)
- Dual-beam spectrophotometer with glass or quartz 1 cm cuvettes

Method

1. Wash the cell monolayers with PBS and add 1.5 ml of Emulgen buffer.
2. Scrape the cells off the plastic and homogenize them by rapid pipetting through a narrow-bore glass pipette.
3. Add 3–4 crystals of sodium dithionite to the homogenate and stir.
4. Distribute the homogenate equally between two matched cuvettes and place them in a dual-beam spectrophotometer.
5. Scan the cuvettes between 400 and 500 nm to establish a flat base line with no significant absorbance.
6. Remove one cuvette and, in a fume hood, bubble CO through it for 30 sec.
7. Replace the cuvette in the spectrophotometer and record the difference spectrum between 400 and 500 nm of the two cuvettes.
8. Divide the height of the peak at 450 nm, i.e. the difference in absorbance measured between 400 and 450 nm, by the molar extinction coefficient of CYP (91 cm^2/mmole) and express the values as nanomole CYP/mg cell protein.

[a] CYP, after reduction by sodium dithionite, is bound to CO and forms a complex with an absorption maximum at 450 nm (8). The total CYP content of the cells is determined by Fe^{2+}–CO versus Fe^{2+} difference spectrophotometry.

Protocol 4. Determination of ethoxyresorufin *o*-deethylase and pentoxyresorufin dealkylase activities[a]

Equipment and reagents

- 10 μM ethoxyresorufin (substrate): prepare a 1 mM solution of 7-ethoxyresorufin (Sigma) in DMSO and divide it into 100 μl aliquots. Evaporate and store at –20°C. For use, redissolve the contents of one aliquot with 20 μl of DMSO, and make up to 10 ml with PBS
- 3 mM salicylamide: prepare a 3 M solution of salicylamide (Sigma) in DMSO and dilute 1:1000 with PBS
- Spectrofluorimeter suitable for reading 96-well plates, e.g. Fluoroskan (Labsystem); filters, 544–584 nm; temperature controlled at 37°C

Protocol 4. *Continued*

- Resorufin (metabolite) standards: prepare a 1 mM solution of resorufin in DMSO and dilute 1:100 with PBS. Use this stock solution to prepare standard solutions ranging from 0.01 to 10 μM
- Cell monolayers in 96-well plates (approximately 1–3 × 10^4 cells/well)

Method

1. Discard the culture medium from the 96-well plate and wash the cells with PBS.
2. Add 50 μl of 3 mM salicylamide to each well, incubate at 37 °C for 1 min, and add 50 μl of 10 μM ethoxyresorufin.
3. Determine the fluorescence of each well every 3 min for 30 min at 37 °C.
4. Prepare the following controls (100 μl/well):
   (a) wells without cells but containing substrate and salicylamide, for the determination of background fluorescence;
   (b) wells without cells but containing resorufin at concentrations ranging from 0.01 to 10 μM, for the determination of the calibration curve;
   (c) wells with cells and containing 0.1 μM resorufin and salicylamide, for verification that the phase II enzymes have been saturated.
5. After 30 min, determine the fluorescence of each well.
6. Plot the calibration curve (fluorescence vs. concentration of resorufin) and use it to calculate the rate of metabolism of the substrate. Express the results as picomoles of resorufin formed/min/mg cell protein.

[a] Ethoxyresorufin is deethylated into the fluorescent metabolite, resorufin, and this reaction is primarily catalysed by CYP 1A. Pentoxyresorufin is also metabolized to resorufin, this reaction being supported mainly by CYP 2B in the rat. In either case, the rate of reaction can be measured by monitoring the increase in fluorescence (9, 10). Salicylamide is added in order to saturate phase II enzyme activities in whole cells and thus prevent the further metabolism of resorufin. To determine pentoxyresorufin dealkylase activity, ethoxyresorufin should be replaced by pentoxyresorufin in this protocol.

Protocol 5. Determination of phenacetin deethylase activity[a]

Equipment and reagents

- Hepatocyte monolayers in 24-well dishes (1–2 × 10^5 cells/well)
- MEM (without phenol red)
- 200 μM phenacetin (substrate): prepare a 40 mM solution of phenacetin (Aldrich) in DMSO. Dilute 1:200 in MEM (without phenol red)
- Paracetamol (metabolite) standards: prepare a 10 mM solution of paracetamol (Sigma) in DMSO and dilute 1:200 with distilled water. Use this stock solution to prepare standard solutions ranging from 2.5 to 50 μM

- 100 μM 3-acetamidophenol (internal standard): prepare a 10 mM solution of 3-acetamidophenol (Aldrich) in methanol and dilute 1:100 with water
- HPLC eluent: 2 mM sodium acetate in 0.75% (v/v) triethylamine:acetonitrile (97:3)
- HPLC apparatus equipped with an automatic injector and a UV detector; wavelength, 250 nm; column, Nucleosil C 18, dp 3 μm, 10 cm (Shandon); flow rate, 1 ml/min

Method

1. Wash the cells twice with MEM.
2. Add 400 μl of substrate solution to each well being tested. Also prepare the following controls (400 μl/well):
   (a) wells without cells but containing substrate, for the determination of the extent of substrate degradation;
   (b) wells with cells but without substrate, for the determination of any analytical interference.
3. Incubate the 24-well plates for 20 h at 37 °C.
4. Collect the media from the wells and store it at –20 °C until required for analysis.
5. After thawing the media, add 10 μl of internal standard solution to 90 μl of each sample or metabolite standard.
6. Inject 50 μl of each sample and metabolite standard into the HPLC apparatus.[b]
7. Calculate the peak areas corresponding to the substrate and metabolite and divide by the peak area of the internal standard. Construct a calibration graph for the metabolite and verify that no more than 60–70% of the substrate has been biotransformed. Express the results as nanomoles of metabolite formed/h/mg cell protein.

[a] Deethylation of phenacetin results in the formation of paracetamol. This reaction is catalysed by CYP 1A2 in humans (11).
[b] Analysis duration for each sample is approximately 15 min. Compounds are eluted in the following order: metabolite, internal standard, and substrate.

Protocol 6. Determination of mephenytoin hydroxylase activity[a]

Equipment and reagents

- 100 μM 5–5 diphenylhydantoin (internal standard): prepare a 10 μM solution of 5–5 diphenylhydantoin (Aldrich) in DMSO and dilute 1:100 with distilled water
- 4-Hydroxymephenytoin (metabolite) standards: prepare a 10 μM solution of 4-hydroxymephenytoin (Sandoz) in DMSO. Use this stock solution to prepare standard solutions ranging from 1–10 μM
- HPLC eluents: 5 mM sodium acetate (adjusted to pH 4.0 with sulfuric acid) and acetonitrile (97:3)
- HPLC apparatus equipped with an automatic injector and a UV detector; wavelength, 210 nm; column, Lichrospher 100 RP, dp 18.5 μm, 12.5 cm (Merck); flow rate = 1.5 ml/min

Protocol 6. *Continued*

- Hepatocyte monolayers in 24-well dishes ($1–2 \times 10^5$ cells/well)
- MEM (without phenol red)
- 200 mM mephenytoin (substrate): prepare a 40 mM solution of mephenytoin (Sandoz) in DMSO. Dilute 1:200 in MEM (without phenol red)

Method

1. Wash the cells twice with MEM.
2. Add 400 µl of substrate solution to each well being tested. Also prepare the following controls (400 µl/well):
   (a) wells without cells but containing substrate, for the determination of the extent of substrate degradation;
   (b) wells with cells but without substrate, for the determination of any analytical interference.
3. Incubate the 24-well plates for 20 h at 37 °C.
4. Collect the media from the wells and store it at –20 °C until required for analysis.
5. After thawing the media, add 10 µl of internal standard solution to 90 µl of each sample or metabolite standard.
6. Inject 50 µl of each sample or metabolite standard into the HPLC apparatus.[b]
7. Calculate the peak areas corresponding to the substrate and metabolite and divide by the peak area of the internal standard. Construct a calibration graph for the metabolite and verify that no more than 60–70% of the substrate has been biotransformed. Express the results as nanomoles of metabolite formed/h/mg cell protein.

[a] Mephenytoin 4-hydroxylase is a prototype of genetic polymorphism in oxidative drug metabolism in humans and is supported by CYP 2C19. This activity is measured by HPLC (12).
[b] Analysis duration for each sample is approximately 21 min. Compounds are eluted in the following order: metabolite, substrate and internal standard.

Protocol 7. Determination of dextromethorphan demethylase activity[a]

Equipment and reagents

- Hepatocyte monolayers in 24-well dishes ($1–2 \times 10^5$ cells/well)
- MEM (without phenol red)
- 100 µM dextromethorphan bromohydrite (substrate): prepare a 20 mM solution of dextromethorphan bromohydrite in DMSO. Dilute 1:200 in MEM (without phenol red)
- Dextrorphan tartrate (metabolite) standards: prepare a 10 mM solution of dextrorphan tartrate in DMSO. Use this stock solution to prepare standard solutions in MEM ranging from 1 to 10 µM
- HPLC eluents: 20 mM sodium perchlorate (adjusted to pH 2.3 with perchloric acid) and acetonitrile (73:27)

- 100 μM thebaine (internal standard): prepare a 10 mM solution of thebaine in DMSO and dilute 1 100 in distilled water
- HPLC apparatus equipped with an automatic injector and a UV detector;[b] wavelength, 214 nm; column, Spherisorb C18, dp 5 μm, 15 cm (Shandon); flow rate = 1 ml/min

Method

1. Wash the cells twice with MEM.
2. Add 400 μl of substrate solution to each well being tested. Also prepare the following controls (400 μl/well):
   (a) wells without cells but containing substrate, for the determination of the extent of substrate degradation
   (b) wells with cells but without substrate, for the determination of any analytical interference.
3. Incubate the 24-well plates for 20 h at 37 °C.
4. Collect the media from the wells and store it at –20 °C until required for analysis.
5. After thawing the media, add 10 μl of internal standard solution to 90 μl of each sample or metabolite standard.
6. Inject 50 μl of each sample or metabolite standard into the HPLC apparatus.[c]
7. Calculate the peak areas corresponding to the substrate and metabolite and divide by the peak area of the internal standard. Construct a calibration graph for the metabolite and verify that no more than 60–70% of the substrate has been biotransformed. Express the results as nanomoles of metabolite formed/h/mg cell protein.

[a] Dextromethorphan is a prototype substrate of the genetic deficiency of oxidative drug metabolism in humans known as debrisoquine sparteine type polymorphism. Demethylation of dextromethorphan it is supported by CYP 2D6. This enzyme activity is measured by HPLC (13).
[b] Dextromethorphan can be detected using fluorescence (excitation wavelength, 270 nm; emission wavelength, 312 nm)
[c] Analysis duration for each sample is approximately 25 min. Compounds are eluted in the following order: metabolite, internal standard, and substrate.

Protocol 8. Determination of chlorzoxazone 6-hydroxylase activity[a]

Equipment and reagents

- Cell monolayers in 24-well plates (1–2 × 10^5 cells/well) or in 35 mm diameter dishes (1–1.5 × 10^6 cells/dish)
- 300 μM chlorzoxazone (substrate): dilute 50 mg chlorzoxazone (Sigma) in 1 ml of DMSO and dilute 1:1000 in MEM (without phenol red)
- MEM (without phenol red)
- 6-Hydroxychlorzoxazone (metabolite) standards: prepare a 10 mM solution of 6-hydroxychlorzoxazone in DMSO. Use this stock solution to prepare standard solutions in MEM ranging from 1 to 10 μM

Protocol 8. *Continued*

- 100 μM 5-fluoro-2-benzoxazolone (internal standard): prepare a 10 mM solution of 5-fluoro-2-benzoxazolone in DMSO and dilute 1:100 with water
- HPLC eluents: Solvent A, 0.5% (v/v) glacial acetic acid in water; Solvent B, acetonitrile
- HPLC apparatus equipped with an automatic injector and a UV detector; wavelength, 287 nm; column, Nucleosil C18, 10 × 4.6 mm or 25 × 4.6 mm (Shandon); flow rate = 1 ml/min

Method

1. Wash the cells twice with MEM.
2. Add the substrate solution to the cells (0.5 ml/well or 1 ml/dish). Also prepare the following controls using the same volumes:
   (a) wells or dishes without cells but containing substrate, for the determination of the extent of substrate degradation;
   (b) wells or dishes with cells but without substrate, for the determination of any analytical interference;
3. Incubate the cells for a maximum of 20 h depending on the rate of metabolism.
4. Collect the media and cells and store at –80°C until required for analysis.
5. After thawing the samples, add 10 μl of internal standard to 90 μl of each sample or metabolite standard.
6. Inject 40 μl of each sample or metabolite standard into the HPLC apparatus and run the following HPLC gradient[b]:

Time (min)	**Solvent A**	**Solvent B**
0	80	20
12	80	20
15	30	70
18	30	70
23	10	90
23.1	80	20
35	80	20

7. Calculate the peak areas corresponding to the substrate and metabolite and divide by the peak area of the internal standard. Construct a calibration graph for the metabolite and verify that no more than 60–70% of the substrate has been biotransformed. Express the results as nanomoles of metabolite formed/h/mg cell protein.[c]

[a] Chlorzoxazone is converted to 6-hydroxychlorzoxazone primarily by CYP 2E1 (14) although CYP 1A is also involved (15). Chlorzoxazone hydroxylation is measured by HPLC (14).
[b] Compounds are eluted in the following order: metabolite, internal standard, and substrate.
[c] The identity of the 6-hydroxychlorzoxazone peak can be confirmed by its chromatographic retention time, UV absorbance, and fluorescence (excitation wavelength, 290 nm; emission wavelength, 326 nm).

Protocol 9. Determination of nifedipine oxidase activity[a]

Equipment and reagents

- Hepatocyte monolayers in 24-well dishes ($1–2 \times 10^5$ cells/well)
- MEM (without phenol red)
- 200 μM nifedipine (substrate): prepare a 40 mM solution of nifedipine (Bayer) in DMSO. Dilute 1:200 in MEM (without phenol red). Prepare only as required as it is light-sensitive
- HPLC apparatus equipped with an automatic injector and a UV detector; wavelength, 240 nm; column, Lichrospher C18, dp 5 μm, 15 cm (Merck); flow rate = 1 ml/min
- Nifedipine dihydropyridine (metabolite) standards: prepare a 50 mM solution of nifedipin dihydropyridine (Bayer) in DMSO. Use this stock solution to prepare standard solutions in MEM ranging from 2.5 to 50 μM
- 100 μM oxodipine (internal standard): prepare a 10 mM solution of oxodipine in DMSO and dilute 1:100 in distilled water
- HPLC eluents: a mixture (63:37) of 5 mM Tris (adjusted to pH 7.5 with sulfuric acid) and acetonitrile

Method

1. Wash the cells twice with MEM.
2. Add 400 μl of substrate solution to each well being tested. Also prepare the following controls (400 μl/well):
   (a) wells without cells but containing substrate, for the determination of the extent of substrate degradation;
   (b) wells with cells but without substrate, for the determination of any analytical interference.
3. Incubate the 24-well plates for 2 h at 37 °C.
4. Collect the media from the wells and store it at –20 °C until required for analysis.
5. After thawing the media, add 10 μl of internal standard solution to 90 μl of each sample or metabolite standard.
6. Inject 50 μl of each sample or metabolite standard into the HPLC apparatus.[b]
7. Calculate the peak areas corresponding to the substrate and metabolite and divide by the peak area of the internal standard. Construct a calibration graph for the metabolite and verify that no more than 60–70% of the substrate has been biotransformed. Express the results as nanomoles of metabolite formed/h/mg cell protein.

[a] Nifedipine is oxidized on the dihydropyridine nucleus to form a pyridyl metabolite. The reaction is catalysed by CYP 3A in humans (16). The metabolite is further slowly transformed.
[b] Analysis duration for each sample is approximately 20 min. Compounds are eluted in the following order: metabolite, substrate, and internal standard.

Protocol 10. Determination of lauric acid hydroxylase activity[a]

Equipment and reagents

- Cell monolayers in 24-well plates ($1–2 \times 10^5$ cells/well)
- PBS
- Tris buffer: 50 mM Tris (pH 7.4) with 1% (w/v) BSA
- 4 M HCl
- 300 μM [^{14}C]lauric acid (substrate): prepare 30 mM solutions of [^{14}C]lauric acid (specific activity 2 MBq/μmole; Amersham) and non-labelled lauric acid (Aldrich) in DMSO. The radioactive solution should produce approximately 40×10^6 d.p.m./ml. Mix the solutions together in equal amounts and dilute 1:100 in Tris buffer
- 10 mM NADPH (cofactor) in Tris buffer
- Extraction solution: ethyl ether
- Ethyl acetate and methanol
- TLC eluent: hexane:chloroform:acetic acid (26:30:2.4)
- TLC chamber and plates (20 × 20 cm; Silica 60; Sigma)
- Autoradiography film (Hyperfilm β-max; Amersham)
- Scintillation counter and scintillation cocktail
- Sonicator
- Vortex mixer
- Fume hood

Method

1. Wash the cells twice with PBS.[b]
2. Add 95 μl of Tris buffer to each well and lyse the cells by sonication.
3. Mix 80 μl of each cell lysate with 10 μl of substrate. Incubate the lysates at 37 °C for 5 min. Also incubate controls containing 80 μl of Tris buffer and 10 μl of substrate only.
4. Add 10 μl of cofactor, stir and reincubate at 37 °C for 15 min. Also incubate controls containing cell lysate and substrate but no cofactor.

   NB Perform steps 5–9 in a fume hood.
5. Acidify each of the samples with 100 μl of 4M HCl and add 2 ml of extraction solution. Vortex cautiously.
6. Centrifuge the samples at 120 *g* for 5 min and collect the organic phase. Evaporate off the ethyl ether.
7. Solubilize each sample with 30 μl of ethyl acetate.
8. Spot the samples on the TLC plate to give approximately 40 000 d.p.m./lane.
9. Place the plate in a TLC chamber containing eluent. Allow the eluent to travel about 18 cm beyond the spotted samples. Dry the plate.
10. Cover the plate with autoradiography film. Develop the film after keeping it in the dark for 4–5 days at −80 °C.
11. Scrape the spots visualized by the film, solubilize in 1 ml of methanol and add to the scintillation cocktail. Count the samples in the scintillation counter.[c]
12. Express the results as nanomoles of metabolite formed/h/mg cell

protein, taking into account the dilution factor of the ^{14}C-labelled substrate. Verify that no more than 60–70% of the substrate has been biotransformed.

[a] Lauric acid is hydroxylated at position 11. The reaction is supported by CYP 4A. A 12-hydroxy metabolite is also found with comparable chromatographic properties. Lauric acid hydroxylation is measured by TLC (17).
[b] Dried monolayers can be stored at −80 °C until required.
[c] Over 90% of the radioactivity must be extracted.

Protocol 11. Determination of the glucuronidation of 4-methylumbelliferone (18)

Equipment and reagents

- Cell monolayers in 24-well plates (1–2 × 10^5 cells/well)
- PBS
- 96-well plates
- 4-Methylumbelliferone (substrate): prepare an 80 mM solution of 4-methylumbelliferone (Sigma) in DMSO and dilute in Tris buffer to produce a range of concentrations between 1 and 400 μM
- 36 mM UDPGA in Tris buffer
- 0.36% (w/v) Triton X100 in Tris buffer
- Spectrofluorimeter suitable for reading 96-well plates, e.g. Fluoroskan (Labsystem); filters, 355–460 mm; temperature controlled at 37 °C
- Sonicator
- Tris buffer: 50 mM Tris (pH 7.4) with 1 mM $MgCl_2$

Method

1. Wash the cells twice with PBS.[a]
2. Add 100 μl of Tris buffer to each well and lyse the cells by sonication.
3. In a 96-well plate, mix 75 μl of each cell lysate with 5 μl of 0.36% Triton X100, 10 μl of 36 mM UDPGA and 90 μl of 400 μM substrate. Also prepare the following controls (180 μl/well):
   (a) wells without lysate but containing Tris buffer, substrate, Triton X100, and UDPGA, to measure any degradation of the substrate during the reaction period;[b]
   (b) wells with lysate, substrate, and Triton X100 but without UDPGA, to measure any metabolism of the substrate other than glucuronidation;[b]
   (c) wells containing 4-methylumbelliferone standards only (1–200 μM), for the determination of the calibration graph.
4. Measure the fluorescence of all the wells every 2 min for 30 min at 37 °C.
5. Construct a calibration graph of substrate concentration versus fluorescence and use it to determine the rate of glucuronidation in the samples. Express the data as nanomoles of 4-methylumbelliferone glucuronidated/min/mg cell protein.

[a] Dried monolayers can be stored at −80 °C until required.
[b] Little change in fluorescence should occur throughout the 30 min incubation.

Protocol 12. Determination of the sulfation and glucuronidation of paracetamol (19, 20)

Equipment and reagents

- Hepatocyte monolayers in 24-well plates ($1-2 \times 10^5$ cells/well)
- MEM (without phenol red)
- 200 μM paracetamol (substrate): prepare a 40 mM solution of paracetamol (Sigma) in DMSO. Dilute 1:200 in MEM (without phenol red)
- 10 μM paracetamol metabolites: prepare a 10 mM solution of paracetamol sulfate and glucuronidate in DMSO and dilute 1:1000 in distilled water
- 24-well plates
- 100 μM 3-acetamidophenol (internal standard): prepare a 10 mM solution of 3-acetamidophenol (Aldrich) in methanol and dilute 1:100 in distilled water
- HPLC eluents: 2 mM sodium acetate in 0.75% (v/v) triethylamine:acetonitrile (97:3)
- HPLC apparatus equipped with an automatic injector and a UV detector; wavelength, 250 nm; column, Nucleosil C18, dp 3 μm, 10 cm (Shandon); flow rate = 1 ml/min

Method

1. Wash the cells twice with MEM.
2. Add 400 μl of substrate to each well being tested. Also prepare the following controls (400 μl/well):
   (a) wells without cells but containing substrate, for the determination of the extent of substrate degradation;
   (b) wells with cells but without substrate, for the determination of any analytical interference.
3. Incubate the 24-well plates for 20 h at 37 °C.
4. Collect the media from the wells and store it at –20 °C until required for analysis.
5. After thawing the media, add 10 μl of internal standard solution to 90 μl of each sample or metabolite standard.
6. Inject 50 μl of each sample or metabolite standard into the HPLC apparatus.[a]
7. Calculate the peak areas corresponding to the substrate and metabolite and divide by the peak area of the internal standard. Verify that no more than 60–70% of the substrate has been biotransformed. Express the results as nanomoles of metabolite formed/h/mg cell protein.

[a] Analysis duration for each sample is approximately 15 min. Compounds are eluted in the following order: glucuronidate, substrate, sulfate, and internal standard. In some analyses, especially if the substrate concentration is raised, e.g. 1 mM, a peak corresponding to a glutathione conjugate is also found.

Protocol 13. Determination of glutathione-*S*-transferase (GST) activities[a]

Equipment and reagents

- 1×10^6 cells in a 20 mm diameter dish
- Ice-cold PBS
- Phosphate buffer: 11 mM KH_2PO_4, 11 mM K_2HPO_4, 50 μM phenylmethane-sulfonylfluoride, and 1 mM EDTA (pH 7.0)
- Spectrophotometer: for wavelength see *Table 1*; temperature maintained at 30 °C
- Substrate and GSH: dissolve double the final desired concentrations of substrate and GSH in 100 mM potassium phosphate buffer (see *Table 1* for final concentrations and pH)
- Sonicator
- Centrifuge

Method

1. Wash the cells with PBS.
2. Add phosphate buffer to the cells and scrape them off the dish.
3. Lyse the cells by sonication and centrifuge the lysate at 4500 *g* for 10 min at 4 °C. Immediately store the supernatant at −80 °C until required.
4. Add 100 μl of supernatant[b] to 450 μl of substrate and 450 μl of GSH (prewarmed in the spectrophotometer).
5. Record the change in absorbance of the reaction mixture for 3 min at 30 °C.[c]
6. Determine the enzyme activity (IU/mg cell protein[d]) using the following equation:

$$\text{Activity (IU/mg protein)} = \frac{\Delta\text{OD/min} \times \text{reaction vol. (ml)}}{\varepsilon \times \text{sample vol. (ml)}} \times \frac{1}{[\text{protein (mg)}]}$$

[a] GST isozymes are categorized into one of four classes (α, μ, π and θ) containing one (π) to several members (α) (21, 22). They are able to catalyse the conjugation of glutathione (GSH) to various substrates. The most frequently used substrates for determining GST activity are 1-chloro-2-4-dinitrobenzene, 1,2-dichloro-4-nitrobenzene and *trans*-4-phenyl-3-butene-2-one.
[b] Use more supernatant if GST activity is likely to be low.
[c] For the initial 3 min of the reaction, the change in absorbance is a linear function of the enzyme concentration as long as the rate of absorbance change is limited to < 0.05/min.
[d] A unit of enzyme activity is defined as the amount of enzyme that will catalyse the formation of 1 μmol of metabolite/min at 30 °C.

Table 1. Conditions for spectrophotometric assays of GST activities (23)

Substrate	Substrate (mM)	GSH (mM)	pH	Wave length (nm)	ε[a] (nM/cm)
1,2-dichloro-4-nitrobenzene	1.0	5.0	7.5	345	8.5
1-chloro-2,4-nitrobenzene	1.0	1.0	6.5	340	9.6
Trans-4-phenyl-3-buten-2-one	0.05	0.25	6.5	290	−24.8
Ethacrynic acid	0.2	0.25	6.5	270	5.0
δ^5-androstene-3,17-dione	0.068	0.1	8.5	248	16.1

[a] ε, Molar extinction coefficient.

Protocol 14. Determination of the *N*-acetylation of procainamide (24)

Equipment and reagents

- Cell monolayers in 24-well plates (1–2 × 10^5 cells/well)
- MEM (without Phenol Red)
- 200 μM procainamide (substrate): prepare a 40 mM solution of procainamide (Aldrich) in DMSO. Dilute 1:200 in MEM (without phenol red)
- *N*-acetylprocainamide (metabolite) standards: prepare a 10 mM solution of *N*-acetylprocainamide (Aldrich) in DMSO. Use this stock solution to prepare standard solutions in MEM ranging from 2 to 20 μM
- 100 μM *N*-propionyl procainamide (internal standard): prepare a 10 mM solution of *N*-propionyl procainamide (Aldrich) in DMSO and dilute 1:100 in distilled water
- HPLC eluents: 10 mM sodium acetate in 0.05% (v/v) triethylamine:acetonitrile (88:12)
- HPLC apparatus equipped with an automatic injector and a UV detector; wavelength, 280 nm; column, Spherisorb C8, dp 3 μm, 15 cm (Merck); flow rate = 1 ml/min

Method

1. Wash the cells twice with MEM.
2. Add 400 μl of substrate solution to each well being tested. Also prepare the following controls (400 μl/well):
   (a) wells without cells but containing substrate, for the determination of the extent of substrate degradation;
   (b) wells with cells but without substrate, for the determination of any analytical interference.
3. Incubate the 24-well plates for 20 h at 37 °C.
4. Collect the media from the wells and store it at −20 °C until required for analysis.
5. After thawing the media, add 10 μl of internal standard solution to 90 μl of each sample or metabolite standard.

6. Inject 50 μl of each sample or metabolite standard into the HPLC apparatus.[a]
7. Calculate the peak areas corresponding to the substrate and metabolite and divide by the peak area of the internal standard. Construct a calibration graph for the metabolite and verify that no more than 60–70% of the substrate has been biotransformed. Express the results as nanomoles of metabolite formed/h/mg cell protein.

[a] Analysis duration for each sample is approximately 15 min. Compounds are eluted in the following order: substrate, metabolite, and internal standard.

5. Conclusions

Cell cultures are widely used in pharmacotoxicological research. However, because of the major role played by the liver in xenobiotic metabolism, the isolated hepatocyte is the most frequently used cell type for metabolism and toxicity studies. Since marked species differences exist between the rates and the routes of drug metabolism, particularly between laboratory animals and humans, the use of human hepatocytes *in vitro* is of major interest. These cells retain both phase I and phase II enzyme activities and can respond to

Table 2. Drug metabolizing enzyme activities in human hepatocytes[a]

Substrate (reaction type)	**Enzyme activity**[b]		
	Min–max	**Mean**	***n***[c]
Ethoxyresorufin[d] (deethylation)	0.2–8	3.0	19
Phenacetin[e] (deethylation)	0.1–25	4.7	25
Pentoxyresorufin[d] (dealkylation)	0.1–5	0.7	19
Mephenytoin[e] (hydroxylation)	0.1–2	0.7	6
Dextromethorphan[e] (demethylation)	0.1–2	0.5	6
Nifedipin[e] (oxidation)	0.5–13	5.3	8
Lauric acid[e] (hydroxylation)	0.3–2	0.8	7
Paracetamol[e]			
—(glucuronidation)	0.3–16	4.1	26
—(sulfation)	0.1–14	3.6	27
1-Chloro-2,4-dinitrobenzene[f] (GSH conjugation)	0.05–0.5	0.2	14
Procainamide[e] (*N*-acetylation)	0.1–7	1.1	32

[a] Unpublished data obtained by C. Chesné.
[b] Enzyme activities were measured 16–48 h after hepatocyte seeding. The minimum, maximum, and mean values are displayed.
[c] *n* = number of tested cell populations from different donors.
[d] Values expressed as picomoles of metabolite formed/min/mg cell protein.
[e] Values expressed as nanomoles of metabolite formed/h/mg cell protein.
[f] Values expressed as IU/mg cell protein.

Table 3. Drug metabolizing enzyme activities in human keratinocytes[a]

Substrate (reaction)	Enzyme activity
Ethoxyresorufin (deethylation)	0.35[b]
Phenacetin (deethylation)	0.10[c]
Paracetamol (glucuronidation)	0.08[c]
Procainamide (*N*-acetylation)	0.19[c]

[a] Enzyme activities were measured after 7 days culture when they were at their peak; mean of 3 or 4 independent experiments (data from ref. 29).
[b] Values expressed as picomoles of metabolite formed/min/mg cell protein.
[c] Values expressed as nanomoles of metabolite formed/h/mg cell protein.

enzyme inducers for at least a few days (3, 7, 25, 26). However, their availability is erratic and unpredictable, and, in addition, major individual variations may exist in the levels of drug metabolizing enzyme activities (see *Table 2*). Therefore, it is essential to determine the functional capacities of each isolated human hepatocyte population and for any study to analyse different cell populations. Since hepatocytes can be entrapped in an alginate gel and cryopreserved (27, 28) it has become possible to perform drug metabolism or toxicity studies on liver cells from various animal species and humans. Cryopreserved immobilized parenchymal cells are now marketed as LIVERBEADS[R] (28).

As shown in this chapter, most protocols for measuring drug metabolizing enzymes require a limited number of cells and can be carried out on whole living cells. These protocols can be employed not only for isolated hepatocytes but also for immortalized and transformed hepatocytes, and for non-hepatic cells which express some drug metabolizing enzyme activities, e.g. keratinocytes (see *Table 3*; 29).

Acknowledgements

We are grateful to Drs C. Guyomard, F. Morel, and Z. Abdel-Razzak for their help in the preparation of this chapter and to Mrs A. Vannier for typing it.

References

1. Seglen, P. O. (1975). *Meth. Cell Biol.*, **13**, 29.
2. Berry, M. N., Edwards, A. M., and Barritt, G. J. (1991). *Laboratory techniques in biochemistry and molecular biology*. Vol. 21. Elsevier, Amsterdam.
3. Guguen-Guillouzo, C. (1992). In *Culture of epithelial cells* (ed. R. I. Freshney), Vol. 1, pp. 197–223. Alan R. Liss, Inc., Glasgow.
4. Guguen-Guillouzo, C., Clément, B., Baffet, G., Beaumont, C., Morel-Chany, E., Glaise, D., and Guillouzo, A. (1983). *Exp. Cell Res.*, **143**, 47.

5. Blaauboer, B. K. J., Boobis, A. E., Castell, J. V., Coecke, S., Groothuis, G. M. M., Guillouzo, A., Hall, T. J., Hawksworth, G. M., Lorenzon, G., Miltenburger, H. G., Rogiers, V., Skett, P., Villa, P., and Wiebel, F. J. (1994). *ATLA*, **22**, 231.
6. Guillouzo, A., Morel, F., Ratanasavanh, D., Chesné, C., and Guguen-Guillouzo, C. (1990). *Toxicol. In vitro*, **4**, 415.
7. Williams, G. M., Weisburger, E. K., and Weisburger, J. M. (1971). *Exp. Cell Res.*, **69**, 106.
8. Omura, T. and Sato, R. (1964). *J. Biol. Chem.*, **239**, 2370.
9. Burke, M. D. and Hallman, H. (1978). *Biochem. Pharmacol.*, **27**, 1539.
10. Lubert, R. A., Mayer, R. J., Cameron, J. W., Nims, R. W., Burke, M. D., Wolff, T., and Guengerich, F. P. (1985). *Arch. Biochem. Biophys.*, **238**, 43.
11. Dislerath, L. M., Reilly, P. E., Martin, M. V., Davis, G. G., Wilkinson, G. R., and Guengerich, F. P. (1985). *J. Biol. Chem.*, **260**, 9057.
12. Shimada, T., Misono, S. K., and Guengerich, F. P. (1986). *J. Biol. Chem.*, **261**, 909.
13. Kronbach, T., Mathys, D., Gut, J., Catin, T., and Meyer, U. A. (1987). *Anal. Biochem.*, **162**, 24.
14. Peter, R., Bocker, R., Beaune, P., Iwasaki, M., Guengerich, F. P., and Yang, C. S. (1990). *Chem. Res. Toxicol.*, **3**, 566.
15. Carrière, V., Goasduff, T., Ratanasavanh, D., Morel, F., Gautier, J. C., Guillouzo, A., Beaune, P., and Berthou, F. (1993). *Chem. Res. Toxicol.*, **6**, 852.
16. Guengerich, F. P., Martin, M. V., Beaune, P. H., Kremers, P., Wolf, T., and Waxman, D. J. (1986). *J. Biol. Chem.*, **261**, 5051.
17. Parker, G. L. and Orton, T. C. (1980). In *Biochemistry, biophysics and regulation of cytochrome P-450* (ed. J. A. Gustafsson, J. C. Carlstedt-Duke, A. Mode and J. Rafter). Elsevier, North-Holland.
18. Bock, K. W. (1976). *Anal. Biochem.*, **72**, 248.
19. Hart, S. J., Tontodonati, R., and Calder, I. C. (1981). *J. Chromat.*, **225**, 387.
20. Moldeus, P. (1978). *Biochem. Pharmacol.*, **27**, 2859.
21. Jacoby, W. B., Ketterer, B., and Mannervik, B. (1984). *Biochem. Pharmacol.*, **33**, 2539.
22. Meyer, D. J., Coles, B., Pemble, S. E., Gilmore, K. S., Fraser, G. M., and Ketterer, B. (1991). *Biochem. J.*, **274**, 409.
23. Habig, W. H. and Jakoby, W. B. (1981). In *Methods in enzymology* (ed. W. B. Jacoby) Vol. 77, pp. 398. Academic Press, London.
24. Rocco, R. M., Abbott, D. C., Giese, R. W., and Karger, B. L. (1977). *Clin. Chem.*, **23**, 705.
25. Morel, F., Beaune, P., Ratanasavanh, D., Flinois, J. P., Yang, G. S., Guengerich, F. P., and Guillouzo, A. (1990). *Eur. J. Biochem.*, **191**, 437.
26. Morel, F., Fardel, O., Meyer, D. J., Langouët, S., Gilmore, K. S., Meunier, B., Tu, C. P. D., Kensler, T. W., Ketterer, B., and Guillouzo, A. (1993). *Cancer Res.*, **53**, 230
27. Frémond, B., Malandain, C., Guyomard, C., Chesné, C., Guillouzo, A., and Campion J. P. (1993). *Cell Transplant.*, **2**, 453.
28. Guyomard, C., Chesné, C., and Guillouzo, A. (1994). *Cell Biol. Toxicol.*, **10**, 445.
29. Hirel, B., Chesné, C., Pailheret, J. P., and Guillouzo, A. (1995). *Toxicol.* In vitro **9**, 49.

5

Culture and characterization of human endothelial cells

VICTOR W. M. VAN HINSBERGH and RICHARD DRAIJER

1. Introduction

Endothelial cells, the lining cells of all blood vessels, are involved in many physiological and pathophysiological processes, including haemostasis, vasoregulation, inflammation, angiogenesis, and the extravasation of fluid, macromolecules, hormones, and leucocytes. The recognition that endothelial cells are involved in all these processes went in parallel with the ability of investigators to study these cells in culture. Endothelial cell culture was first established in the early 1970s (1, 2). Nowadays, a whole repertoire of molecular and cellular approaches are available to study the activation and metabolic regulation of endothelial cells *in vitro* and *in vivo*. Although one has to realize that endothelial cells from different types of vascular beds have distinct features, much information regarding these cells can be obtained using cell culture methods. This chapter describes the culture and characterization of human umbilical vein and microvascular endothelial cells, and simple techniques to study the permeability of human endothelial cell monolayers *in vitro*.

2. Materials for the isolation and culture of endothelial cells

2.1 Sera

Human serum is prepared from freshly collected blood obtained from healthy donors. The sera of 15–25 subjects are pooled and stored at 4 °C (for up to 3 months) or at –80 °C. Before use the sera are filtered through a 0.45 μm Acrodisc (bottle-top) filter at room temperature. Human serum can be heat-inactivated by incubating for 30 min at 56 °C (i.e. the bottle contents should be at 56 °C for 30 min!). The quality of every batch of serum is tested before use.

Newborn calf serum (NBCS) is purchased from Gibco–BRL or another commercial supplier. It is stored at –20 °C. NBCS has to be heat-inactivated (30 min at 56 °C) before use.

2.2 Preparation of endothelial cell growth factor (ECGF)

A crude preparation of ECGF can be made according to the method described by Maciag *et al.* (ref. 3; see *Protocol 1*). The lyophilized preparation can be stored at 4 °C for 6 months. It is dissolved, for example, in M199 medium supplemented with 20 mM Hepes and 5 U/ml heparin (the latter is optional but it improves the stability of the growth factor), and sterilized by filtration (0.22 µm filter; only once, to prevent loss of material). Heparin (5000 IU/ml) can be purchased from Leo Pharmaceuticals or other commercial suppliers. The crude ECGF preparation can be further purified by ammonium sulfate precipitation and heparin–Sepharose affinity chromatography as described by Burgess *et al.* (4).

Protocol 1. Preparation of crude ECGF from bovine brain according to the method of Maciag *et al.* (3)

Equipment and reagents

- A bovine brain (about 600 g) obtained (aseptically) from a local slaughterhouse. It can be collected in advance and stored at –20 °C if necessary
- 0.1 M NaCl (ice-cold)
- Streptomycin sulfate (powder; Boehringer–Mannheim)
- Dialysis tubing
- Blender
- Centrifuge (e.g. Sorvall RC5) with GSA rotor and 350 ml centrifugation bottles
- Freeze-dryer (e.g. Virtis Freeze Mobile type 12SL) and appropriate size bottles for dry-freezing

Method

1. Start the procedure with the brain at ice-cold temperature.
2. Remove the blood-containing regions, and cut the brain into pieces of about 1 cm^2.
3. Homogenize the pieces of brain for 3 min in 0.1 M NaCl in an ice-cold blender (total 500 ml NaCl solution). Keep the pH at 7.0 during this procedure.[a]
4. Stir the homogenate (pH 7.0) for 2 h at 4 °C.
5. Subsequently, centrifuge the homogenate for 40 min at 13800 *g* and at 4 °C. Recover the supernatant.
6. Add streptomycin sulfate to the supernatant to 0.5% (w/v) final concentration and incubate for at least 1 h (or overnight) at 4 °C to extract lipid material. Check the pH and keep it at 7.0.
7. Centrifuge the mixture at 4 °C for 40 min at 13 800 *g*.
8. Dialyse the supernatant against 0.1 M NaCl overnight at 4 °C.
9. Centrifuge the dialysed solution for 40 min at 13 800 *g* and at 4 °C.

10. Lyophilize the supernatant.

11. Store the lyophilized preparation at 4°C (stability > 6 months).

[a]Check once or twice during homogenization or immediately after homogenization and, if needed, adjust the pH by adding HCl or NaOH.

2.3 Coating tissue culture dishes and coverslips

All dishes and flasks used to culture endothelial cells are tissue culture grade and are coated with fibronectin or gelatin (see *Protocol 2*, part A). Both types of coating give similar results. For immunofluorescence studies it is preferable to grow the cells on glass coverslips because many plastics fluoresce. Round coverslips (14 mm in diameter) can be ordered which fit into 24-Multiwell dishes, or rectangular coverslips can be placed in larger wells. It is advisable to coat the coverslips with cross-linked gelatin (see *Protocol 2*, part B) otherwise the cells may detach after a couple of days in culture.

Protocol 2. Coating tissue culture dishes and coverslips for the culture of human endothelial cells

Equipment and reagents

- PBSA: 150 mM NaCl, 10 mM Na_2HPO_4, and 1.5 mM KH_2PO_4 (pH 7.4)
- 1% (w/v) gelatin (Merck) dissolved in PBSA, sterilized by autoclaving (121°C, 1 bar, 45 min), and stored at room temperature[a]
- 10 μg/ml human fibronectin dissolved at 37°C in M199 medium and sterilized by filtration through a 0.2 μm filter[b]
- 70% and 96% (v/v) ethanol
- 0.5% (w/v) glutaraldehyde in PBSA
- M199 medium (pH 7.4) (ICN-Flow)
- Laminar-flow hood
- Tissue culture dishes (see above)
- Coverslips (see above)
- Sterile forceps

NB Perform these procedures in a laminar flow hood.

A. *Coating tissue culture dishes*

1. Cover the substrate surface of the dish or flask with a thin layer of 1% gelatin or 10 μg/ml fibronectin.
2. Incubate at room temperature for (at least) 30 min.
3. Aspirate the coating solution immediately before seeding the cells.

B. *Coating coverslips*

1. Defat and sterilize the coverslips by washing them in 70% ethanol and then with 96% ethanol in a culture dish.
2. Aspirate the ethanol and wash the coverslips with PBSA.
3. Incubate the coverslips for 45 min in 1% gelatin.
4. Aspirate the gelatin solution and cross-link the adhering gelatin by adding 0.5% glutaraldehyde. Incubate at room temperature for 15 min.

Protocol 2. *Continued*

5. Remove the glutaraldehyde solution and wash the coverslips vigorously with M199 medium (twice).
6. Incubate the coverslips for 15 min at room temperature in M199 medium and subsequently wash again several times with M199 medium.
7. Use the coverslips, or store them dry until use.

[a] No endotoxin was found in our batch of gelatin by Limulus assay.
[b] Fibronectin can be purchased from commercial suppliers or can be prepared from the cryoprecipitate of human plasma obtained from a local blood transfusion service. The fibronectin is purified by gelatin–Sepharose chromatography according to the method of Vuento and Vaheri (5), which yields a mixture of fibronectin and vitronectin. It is dialysed against 10 mM Caps buffer (pH 11.0) supplemented with 1 mM $CaCl_2$ and 150 mM NaCl, and subsequently stored at –80°C in 1 mg aliquots.

2.4 Preparation of DiI-acetylated LDL

Acetylated low-density lipoproteins labelled with the fluorescent dye 3,3′-dioctadecylindocarbocyanine (DiI) are avidly taken up by endothelial cells and macrophages (6). As such, DiI-acetylated LDL uptake is used to identify living endothelial cells during cell sorting. DiI-acetylated LDL can be prepared from human LDL (see *Protocol 3*). LDL are prepared from the blood of a healthy donor by gradient ultracentrifugation according to the method of Redgrave *et al.* (7). The LDL fraction is sliced from the tube, the protein content is determined, and the LDL are dialysed overnight against PBSA at 4°C.

Protocol 3. Preparation of DiI-acetylated LDL from low-density lipoproteins (LDL)

Equipment and reagents

- PBSA (see *Protocol 2*)
- Saturated sodium acetate in water
- M199 medium (with Hanks' salts) supplemented with 20 mM Hepes (pH 7.4) (Flow, cat no. 12–234–54). To this medium add 350 mg/litre sodium bicarbonate, 2 mM L-glutamine, 100 IU/ml penicillin, and 0.10 mg/ml streptomycin (final concentrations)
- Acetic anhydride
- 1% (w/v) BSA in PBSA
- 3 mg/ml DiI: dissolve 3 mg DiI (Molecular Probes, Inc.) into 1 ml DMSO
- Human LDL and lipoprotein-depleted serum (dialysed overnight against M199 medium at 4°C; usually to be obtained in collaboration with a laboratory working on lipoproteins)
- Filters for sterilization (0.45 μm pore size)
- 5 ml syringe filled with Sephadex G-50

A. *Acetylation of LDL (8)*

NB Perform this procedure at 0°C (on ice) and wear safety glasses.

1. Add 2 ml saturated sodium acetate solution to 2 ml of the LDL solution (1–1.5 mg LDL protein/ml).

2. Add, by the addition of small amounts and under continuous stirr. 3–4.5 μl acetic anhydride.
3. Stir for 30 min on a magnetic stirrer.
4. Dialyse the LDL solution against PBSA for 3 h (two renewals of PBSA) and finally overnight against M199 medium.
5. Stabilize the obtained acetylated LDL preparation with 1% albumin or 20% lipoprotein-depleted human serum, and sterilize by filtration through a 0.45 μm filter.

B. *Preparation of DiI-acetylated LDL*

1. Mix 1 ml acetylated-LDL (1 mg/ml) with 2 ml lipoprotein-depleted serum, and add 50 μl of 3 mg/ml DiI.
2. Incubate the mixture for 18 h at 37 °C.
3. Remove unbound DiI from the DiI-acetylated LDL preparation by chromatography on Sephadex G-50.[a] Recover the DiI-acetylated LDL fraction.
4. Filter sterilize the DiI-acetylated LDL preparation and take a sample for protein or cholesterol determination.
5. Add 10% (v/v) lipoprotein-depleted serum (to stabilize the preparation) and store at 4 °C in a dark environment (wrap the tube in aluminium foil).[b]

[a] The large DiI-acetylated LDL particles (M_w about 2×10^6 Da) leave the column immediately after the void volume. A rapid separation can be obtained by centrifuging a 5 ml sterile syringe filled with Sephadex G-50, loading the DiI-acetylated LDL preparation on it, and centrifuging again for 2 min at 500 *g* at room temperature.
[b] During storage some aggregation may occur. Filter before use to remove these aggregates.

2.5 Precautions

The use of human material requires strict safety precautions to be taken. Wear sterile gloves, a laboratory coat, and work in a down-flow sterile hood. Human endothelial cells may be, or may become, infected with viruses or mycoplasma. This not only results in the endothelial cells having altered properties, but also offers the possibility that pathological material may become transferred. Sterilize discarded materials and dishes by autoclaving.

All materials and solutions should be pyrogen-free otherwise endothelial cells will become inflammatory activated. Check the cells regularly in your laboratory for mycoplasma contamination. A convenient assay is the mycoplasma detection system of GenProbe (San Diego, CA) which is based on hybridization with a complementary ^{3}H-labelled DNA probe that recognizes many species of mycoplasma.

The use of amphotericin B (Fungizone; Flow) is sometimes advocated to prevent contamination with yeast or moulds. Although this cannot always be circumvented, for example in establishing primary cultures of foreskin microvascular endothelial cells, its use should be limited and rejected for experiments. Amphotericin B forms complexes with cholesterol in the plasma membrane of the cells which obstruct the pores. Consequently cell properties may become altered in the presence of amphotericin B.

Culture media should be stored in the dark because Hepes may form toxic oxidation products in the light. As streptomycin may also form reactive oxidation products with time, culture media should not be stored at 4°C for longer than a few weeks.

3. Isolation and culture of human endothelial cells

3.1 Isolation and culture of umbilical vein endothelial cells

An umbilical cord usually contains three large blood vessels, of which the vein is the largest. Two arteries are also present which are usually closed by contraction of the vessels. Endothelial cells are generally isolated from the umbilical vein using the methods of Jaffe *et al.* (1) and Gimbrone *et al.* (2) (see *Protocol 4*). It is also possible to isolate endothelial cells from an artery and subsequently from the vein of the same umbilical cord (8).

Protocol 4. Isolation and culture of endothelial cells from human umbilical vein according to the method of Jaffe *et al.* (1) with minor modifications

Equipment and reagents

- CO_2 incubator (water-conditioned)
- Laminar-flow hood (preferably down-flow)
- Inverted phase-contrast microscope
- Laboratory centrifuge
- Kocher scissors, scalpel, forceps, scissors
- Surgical silk (e.g. Ethicon Leinenzwirn 3.5)
- 50 ml sterile plastic tubes (Falcon)
- Clean laboratory coat
- Sterile gloves
- Sterile endotoxin-free pipettes (e.g. Costar)
- 20 ml sterile syringes
- Cannulas: two sterile cannulas to which a short piece of silicon tube is mounted (see *Figure 1A*)
- Fibronectin- or gelatin-coated dishes/flasks (see *Protocol 2*): tissue culture multiwell dishes or 25 cm^2 flasks (Costar, Nunc, etc.)
- Umbilical cord
- Cord buffer: 140 mM NaCl, 4 mM KCl, 11 mM D-glucose, 10 mM Hepes, 100 IU penicillin, and 0.10 mg/ml streptomycin (pH 7.3)
- M199 medium: M199 medium (with Earle's salts) supplemented with 20 mM Hepes (pH 7.4) (ICN-Flow, cat. no. 12–204–54). To this medium add 2.2 g/litre $NaHCO_3$, 2 mM L-glutamine, 100 IU penicillin, and 0.10 mg/ml streptomycin (final concentrations)
- 0.1% collagenase: 0.1% (w/v) type I collagenase (Worthington; CLS II) dissolved in M199 medium[a]
- Culture medium: M199 medium (supplemented with 20 mM Hepes) with 10% human serum, 10% NBCS (heat-inactivated), 150 μg/ml crude ECGF, 5 U/ml heparin, 100 IU/ml penicillin, and 0.10 mg/ml streptomycin (pH 7.4)[b]
- 0.5–1 litre 0.9% NaCl in a sterile beaker

- Solution A: 137 mM NaCl, 5.4 mM KCl, 4.3 mM $NaHCO_3$, 5 mM D-glucose and 0.002% (w/v) phenol red (pH 7.3)
- Trypsin/EDTA: 0.05% (w/v) trypsin, 137 mM NaCl, 5.4 mM KCl, 4.3 mM $NaHCO_3$, 5 mM D-glucose, and 0.67 mM EDTA (pH 7.3)

A. *Collection and storage of umbilical cords*

1. After delivery, cut the umbilical cord from the placenta, and place in ice-cold cord buffer.
2. Store in a refrigerator until use.[c]

B. *Isolation and culture of endothelial cells*

1. The isolation procedure must be performed in a sterile environment, e.g. a down-flow laminar-flow hood. Use sterile gloves and a laboratory coat to prevent infection of the cultures, and to protect yourself.
2. Inspect the umbilical cord for clamped or otherwise damaged areas. Discard these areas. Next, cut a small piece off one end of the cord so that a fresh cut is obtained.
3. Insert the sterile cannula, to which a small tube is connected, into the vein (filled with cord buffer), and fix it tightly in place using surgical silk.
4. Connect a syringe filled with 20–25 ml of cord buffer to the tube joining the cannula. Remove the blood from the vein by slowly flushing the blood vessel and draining the contents into a sterile waste beaker. Repeat if necessary to remove all the blood. Ensure that no air is introduced into the lumen of the vessel.[d]
5. Cut a small piece off the other end of the umbilical cord. Insert the second cannula into the vein and fix it tightly in place using surgical silk. Fill the vein with collagenase solution via the small tube that is joined to the cannula. Clamp the tubes on both sides with Kocher scissors.
6. Incubate the filled (slightly distended) vessel for 20 min at 37 °C in a beaker containing 0.9% NaCl.
7. Remove the clamps and collect the contents of the vein lumen into a 50 ml sterile plastic tube. Flush the vessel with an additional 20 ml of M199 medium.
8. Centrifuge the tube for 5 min at 200 *g* and resuspend the pelleted cells (often a mixture of groups of endothelial cells and red blood cells) in 3–5 ml of culture medium.
9. Seed the cells into fibronectin- or gelatin-coated multiwell dishes (total area 20–30 cm^2) or a similarly coated 25 cm^2 flask.
10. Put the dishes into an incubator (water-conditioned) at 37 °C under an atmosphere of 5% CO_2/95% air.

Protocol 4. *Continued*

11. Wash the adhered cells one or two days after isolation and renew the culture medium (1.5 ml/10 cm^2). Inspect the cells using an inverted phase-contrast microscope.
12. Renew the culture medium and inspect the growth of the cells every 2–3 days.

C. *Serial propagation of endothelial cells*

1. When the cells have become confluent, they can be detached and seeded into new dishes.
2. Wash the cells with solution A and detach them from the culture vessel by coating them with trypsin/EDTA. Observe the cell detachment using a phase-contrast microscope. As soon as the cells start to detach, gently tap the side of the culture vessel to ensure detachment of all the cells.
3. Immediately add serum-containing medium to inhibit any further trypsin activity which will damage the cells.
4. Slowly triturate the cell suspension using a sterile pipette, and transfer the cells into new fibronectin- or gelatin-coated dishes (split ratio 4:1) containing an adequate amount of culture medium.[e]

[a]Different batches of collagenase have different specific activities. Usually a concentration of 0.05–0.1% is sufficient for isolation.
[b]Instead of using a mixture of human serum and NBCS, 20% NBCS can be used. A proper batch of fetal calf serum can also be used, but our experience is that NBCS is more satisfying (see also ref. 9).
[c]Isolation of endothelial cells yields more cells if the cords have been stored at 4°C for at least 6 h. Storage can last as long as 48 h without deterioration of the cell yield.
[d]During this procedure a distention of the blood vessel is often seen. This may occur because of small blood aggregates obstructing the vessel. Press slowly to push the aggregates out of the vessel, but avoid any sudden release of pressure. If this does not work, repeat the procedure at the other end of the vessel (valves may be involved; they act only in one direction). Make sure that you are not swelling the Warthon jelly outside the vessel which occurs when you have put the cannula through the vessel wall.
[e]Umbilical vein endothelial cells can be serially propagated for 30–70 population doublings, but after about 5–7 passages the cells gradually start to increase in size, to grow more slowly, and to lose specific functions (10, 11).

3.2 Isolation and culture of endothelial cells from adult arteries and veins

Endothelial cells can be isolated from segments of adult human arteries and veins, obtained from surgery or autopsy, by a similar method as that used to isolate endothelial cells from umbilical veins (12, 13). Before cannulating the vessel, small branching blood vessels, such as intercostal arteries, are closed by ligation with surgical silk, taking care that the inside of the vessel remains

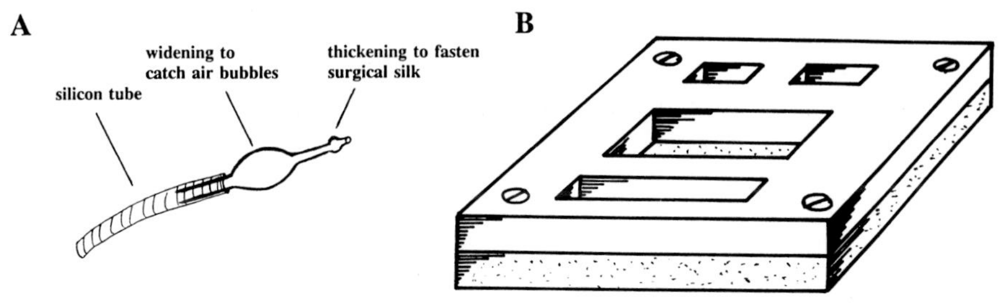

Figure 1. Equipment for the isolation of human endothelial cells. (A) A siliconized glass cannula, two of which are used for the perfusion of an umbilical vein. On one end of the cannula a sterile tube is connected to which syringes can be mounted. (B) Two plates of Perspex, the upper one of which contains a number of slits of different size. A small piece of a blood vessel can be clamped between the two plates at an appropriate position (i.e. with all the edges between the Perspex plates). After screwing the plates together, the slit is filled with collagenase solution for the isolation of endothelial cells.

wet during this procedure. Subsequently, the vessel is cannulated and incubated with collagenase as previously described (see *Protocol 4*) adapting volumes according to the size of the vessel. The cells are seeded and cultured as detailed in *Protocol 4*. It is important to observe the cells after seeding, in particular endothelial cells obtained from adult arteries. These preparations may contain debris (e.g. thrombi and calcium phosphate crystals) and contaminating smooth muscle cells (still elongated immediately after isolation) released from advanced arteriosclerotic lesions. Usually spreading of the endothelial cells is visible 4 h after seeding the cells. At this stage, the cells are washed vigorously to remove the debris and smooth muscle cells which adhere more slowly than the endothelial cells. Although some endothelial cells will also be removed, this step is necessary to prevent contamination of the endothelial cell culture by smooth muscle cells. As an alternative approach, endothelial cells may be selected by FACS (see *Protocol 6*) or by using Dynabeads coated with endothelial specific antibodies.

Serial propagation of endothelial cells from human arteries and veins can be performed for 40–70 population doublings (8, 12, 13). However, whereas human vein endothelial cells keep a normal small diameter morphology for 12–14 passages (split ratio 4:1), endothelial cells isolated from arteries of adult people remain healthy for only 4–7 passages. Thereafter they have a larger diameter and display characteristic signs of senescence, including a retarded growth rate (13).

3.3 Isolation of endothelial cells from small pieces of large blood vessels

When small pieces of blood vessels are obtained from surgery, the opened vessel can be stretched on a sterile Perspex plate with its luminal side on top (prevent drying!). A second sterile Perspex plate with holes of various sizes is placed upon it in such a way that only the endothelial cell-covered surface is exposed in one of the holes (see *Figure 1B*). All the margins should be covered to prevent isolation and contamination by fibroblasts or smooth muscle cells. Subsequently, the hole is filled with a 0.1% collagenase solution and incubated for 20 min at 37 °C. After incubation, the isolated cells are obtained by aspirating with a sterile plastic pipette and gentle washing of the exposed area. The remaining procedure is the same as that detailed in *Protocol 4*.

3.4 Isolation and culture of microvascular endothelial cells from human foreskin

Microvascular endothelial cells can be isolated from various human tissues, including foreskin (14, 15), adult dermis (16), lung (17), and brain (18). In principle, two steps are required to obtain a culture of pure microvascular endothelial cells. Firstly, the cells have to be isolated from the tissue. Secondly, the endothelial cells have to be purified by selection and/or by removal of contaminating cells. The latter can be achieved by FACS or by selective binding of endothelial cells to Dynabeads (beads with an iron core which can be separated by a magnet) to which endothelial specific antibodies are cross-linked. The procedures to isolate and purify human foreskin microvascular endothelial cells (14, 15) are given in *Protocols 5* and *6*.

Human foreskin microvascular endothelial cells grow rather slowly. They can be propagated for about 10 passages (split ratio 3:1) and retain a healthy appearance. Thereafter they often show a senescent morphology. Living cells can be stored in liquid nitrogen (see Section 5.1, *Protocol 8*).

Protocol 5. Isolation and culture of foreskin microvascular endothelial cells according to the method of Davison *et al.* (14) with minor modifications

Equipment and reagents

- CO_2 incubator (water-conditioned)
- Laminar-flow hood (preferably down-flow)
- Inverted phase-contrast microscope
- Laboratory centrifuge
- Sterile rubber plug, 6–8 cm diameter
- Sterile needles
- Sterile forceps
- Sterile scalpel or dermatome
- Clean laboratory coat
- Sterile gloves
- Sterile endotoxin-free pipettes (Costar)
- 50 ml centrifuge tubes (Falcon)
- 20 ml sterile syringes
- Trypsin solution: 0.3% (w/v) trypsin, 1% (w/v) EDTA, 137 mM NaCl, and 5.4 mM D-glucose (pH 7.3)
- Sterile PBSA (see *Protocol 2*)
- Olive oil
- Cord buffer (see *Protocol 4*)
- M199 medium (see *Protocol 4*)
- M199 medium supplemented with 10% (v/v) NBCS (heat inactivated)
- Culture medium: M199 medium (supplemented with 20 mM Hepes) with 20% human serum, 10% NBCS (heat-inactivated), 100 IU/ml penicillin, and 0.10 mg/ml streptomycin (pH 7.4)[a]
- Culture medium supplemented with 2.5 µg/ml amphotericin B
- Trypsin/EDTA (see *Protocol 4*)
- Fibronectin-coated dishes (see *Protocol 2*)

Method

1. Perform the isolation procedure in a sterile environment, e.g. a down-flow laminar-flow hood. Use sterile gloves and a laboratory coat to prevent infection of the cultures, and to protect yourself.
2. Collect the foreskin in cord buffer and store it at 4°C until use. Use it within 24 h.
3. Wash the piece of foreskin with cord buffer or PBSA, and stretch it using needles on to a sterile rubber plug.
4. Put a drop of olive oil on the foreskin and spread it; this facilitates the cutting.
5. Cut very thin slices with a scalpel or dermatome; if possible discard the epidermis.
6. Incubate the pieces in trypsin solution for 30 min at 37°C.
7. Terminate the trypsin activity by adding an equal volume of M199 medium with 10% NBCS.
8. Squeeze the remaining tissue in serum-containing medium to force the cells out. Discard the residual pieces of skin and the epidermis (which floats).
9. Collect the cells in the incubation medium and the fluid obtained after squeezing by centrifugation for 5 min at 200 *g*.
10. Resuspend the cells in 3–5 ml culture medium supplemented with amphotericin B, and seed them into fibronectin-coated dishes.
11. Put the dishes in an incubator (water-conditioned) at 37°C under a 5% CO_2/95% air atmosphere.
12. Wash the cells one or two days after isolation and renew the culture medium (1.5 ml/10 cm^2). Visually inspect the cells by using an inverted phase-contrast microscope.
13. Detach the cells two or three days after isolation by treating them with trypsin/EDTA. Reseed the cells into new fibronectin-coated dishes of the same surface area. This process should remove any epithelial cells present.[b]
14. Renew the culture medium and inspect the cell growth every two to three days.[c]

[a]10 ng/ml basic fibroblast growth factor (human recombinant bFGF; e.g. Boehringer–Mannheim) and 5 U/ml heparin may be added to the culture medium. The addition of ECGF is not advised because the crude preparation also contains a growth factor for melanocytes which may consequently contaminate the cultures. Therefore, the addition of ECGF is postponed until the endothelial cells are purified (see *Protocol 6*).
[b]Epithelial cells attach very firmly and, therefore, remain in the dish after mild trypsinization.
[c]Considerable variation exists between isolations with respect to the yield of endothelial cells and the contamination by other cell types. After the given procedure, endothelial cells may represent between 10 and 60% of the cell population. Hence, additional purification is necessary to prevent overgrowth by fibroblastoid cells (see *Protocol 6*).

Protocol 6. Purification of foreskin microvascular endothelial cells by FACS according to the method of Voyta *et al.* (15)

Equipment and reagents

- Beckton-Dickinson FACS IV cell sorter
- 6 ml sterile tubes (Falcon; cat. no. 2058)
- Fibronectin-coated dishes (see *Protocol 2*)
- Culture medium (see *Protocol 4*)
- 10 μg/ml Dil-acetylated LDL: 10 μg/ml Dil-acetylated LDL (see *Protocol 3*) in culture medium
- Solution A (see *Protocol 4*)
- Trypsin/EDTA (see *Protocol 4*)
- M199 medium: M199 medium (with Hanks' salts) supplemented with 20 mM Hepes (pH 7.4)
- Endothelial cells from *Protocol 5*
- M199 medium with 10% NBCS (heat-inactivated)
- Laboratory equipment for the culture of endothelial cells (see *Protocol 5*)

Method

1. Incubate the cells for 4 h at 37 °C with 10 μg/ml Dil-acetylated LDL.[a]
2. Wash the cells with solution A and detach them with trypsin/EDTA to obtain a single cell suspension.
3. Neutralize the trypsin by washing the cells with M199 medium supplemented with 10% NBCS.
4. Immediately before cell sorting, centrifuge the cells for 5 min at 200 *g* and resuspend them in serum-free M199 medium.
5. Sort the endothelial cells from the other cell types by using a Beckton-Dickinson FACS IV cell sorter with an argon laser. The laser should have an excitation wavelength of 514 nm and emission fluorescence should be above 550 nm.[b] Keep the cells and collection tubes cooled on ice during the whole procedure.
6. Collect the endothelial cells into tubes containing M199 medium supplemented with 10% human serum.
7. After completion of the cell sorting, spin the cells down (5 min at 600 *g*) and wash them once. Resuspend the cells in culture medium and seed them into fibronectin-coated dishes.
8. Renew the culture medium every two or three days. Inspect the growth and purity of the cells by phase-contrast microscopy.[c]

[a]During this period the cells become loaded with Dil-acetylated LDL. This can be checked by fluorescence microscopy (see *Protocol 7*). Endothelial cells and macrophages are strongly labelled by Dil-acetylated LDL. However, only endothelial cells are transferred and recovered during subculturing.

[b]Sampling gates are set using positive and negative cells (a small sample of the cell suspension, or, if a limited number of cells are available, by using umbilical vein endothelial cells and fibroblasts).

[c]Sometimes the purification procedure has to be repeated to obtain a pure endothelial cell culture. In particular, melanocytes stick to endothelial cells during cell sorting. They may become prominent when the endothelial cell culture is kept quiescent at confluency. By selecting simultaneously for Dil-acetylated LDL positive cells and for only single cells, 'hitch-hiking' of contaminating cells can be prevented.

4. Identification of human endothelial cells

4.1 Morphology, ultrastructure, and von Willebrand factor (vWF)

To identify the endothelial nature of the cells obtained in culture, positive and negative criteria are used. *Table 1* summarizes a number of determinants which are used to identify endothelial cells. In addition, other markers, such as smooth muscle α-actin (smooth muscle cells and pericytes) and cytokeratins (epithelial cells and mesothelial cells; see Chapter 2), which are absent in normal endothelial cells, are used to evaluate the presence of contaminating cells in the culture.

Routine evaluation of the quality and growth state of the cultured cells is achieved using an inverted phase-contrast microscope at 100 × magnification. Endothelial cells display a cobblestone morphology at confluence (see *Figure 2A*). Upon prolonged maintenance in the confluent state, some cells acquire a 'sprouting' phenotype and start to infiltrate under the other cells. This disorganization of the monolayer increases with time; it is reduced or delayed by the presence of ECGF (or pure aFGF or bFGF) in the culture medium. When the cells are detached by trypsin/EDTA and transferred to new fibronectin-coated dishes, the cells grow normally again.

Evaluation of human endothelial cells by transmission electron microscopy can be performed by washing the attached cells with 1.5% glutaraldehyde in 100 mM cacodylate buffer (pH 7.4) for 10 min at room temperature. The cells are postfixed with 1% OsO_4 in phosphate buffer with 0.05 M potassium hexacyanoferrate (II) (19). Following dehydration in a graded 70–100% (v/v) ethanol series, the cells are embedded *in situ* in Epon. Ultrathin sections are stained, and visualized using an electron microscope (e.g. Philips EM 300 electron microscope) at 80 kV. Characteristics of endothelial cells include a flat cell shape, multiple small vesicles, a distinctively shaped nucleus, and

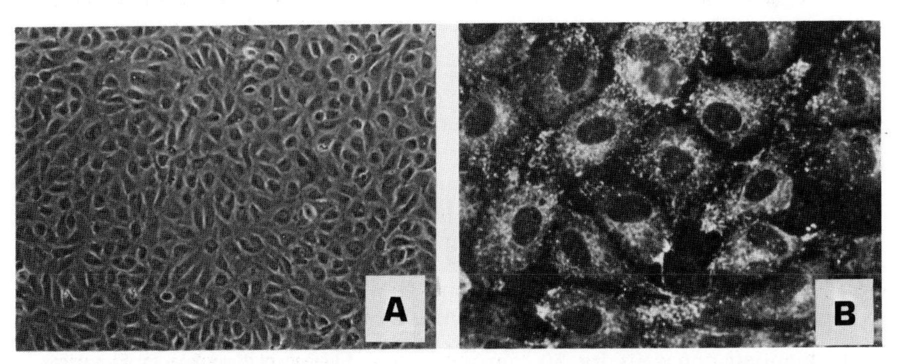

Figure 2. Characterization of human umbilical vein endothelial cells. (A) Phase-contrast photomicrograph. (B) Indirect immunofluorescent staining showing the presence of vWF.

Weibel–Palade bodies (20). Weibel–Palade bodies are storage granules containing vWF and P-selectin (21, 22). They are only encountered in endothelial cells and are found relatively frequently in umbilical vein endothelial cells.

The presence of vWF (factor VIII related antigen) in specific storage organelles in endothelial cells is widely used as a marker for the endothelial nature of cells (see *Figure 2B*). *Protocol 7* describes the visualization of vWF by immunofluorescence microscopy. Von Willebrand factor is only expressed in endothelial cells and in megakaryocytes, and has been encountered in all types of human endothelial cells. However, the absence of vWF has been observed in specific types of animal-derived endothelial cells. For example, pig aorta endothelial cells do not contain vWF, whereas the endothelial cells from the adventitial vessels of the same animal do.

Protocol 7. Determination of the presence of vWF by immunofluorescence microscopy

Equipment and reagents

- Leitz epi-illumination fluorescence microscope
- Endothelial cells grown on gelatin-coated, 14 mm diameter coverslips (see *Protocol 2*)[a]
- Coverglasses and object glasses
- Sterile PBSA (see *Protocol 2*)
- 80% acetone (v/v) in distilled water
- vWF antiserum: rabbit anti-human vWF immunoglobulins (1:80 diluted in 10% pig serum) (AO82; DAKO A/S, Denmark)
- FITC-labelled pig anti-rabbit immunoglobulin: fluorescein isothiocyanate-conjugated swine anti-rabbit immunoglobulins (DAKO immunoglobulins, Denmark) diluted 1:50 in PBSA
- PPD mountant: 100 mg of *p*-phenylenediamine (Sigma) dissolved in 10 ml PBSA and subsequently added to 90 ml glycerol. Final pH is adjusted to approximately 8.0 with 0.5 M carbonate–bicarbonate buffer (pH 9.0)[b]

Method

1. Aspirate the culture medium from the cells.
2. Wash the coverslips with PBSA and fix the cells in 80% acetone for 10 min at 4 °C.
3. Wash the fixed cells with PBSA.
4. Incubate the cells for 30 min at room temperature with vWF antiserum. Use 50 μl/coverslip.[c]
5. Wash the cells three times with PBSA.
6. Incubate the cells for 30 min with 50 μl FITC-labelled pig anti-rabbit immunoglobulin.
7. Wash the cells three times with PBSA.

8. Mount the cells with PPD mountant under glass coverslips and observe them under a fluorescence microscope using an excitation wavelength of 450–490 nm and an emission wavelength of > 515 nm.

[a] Circular coverslips with a diameter of 14 mm fit nicely in a 24-well dish. After fixation in the dish, the coverslips can be taken out by using a hooked needle, and mounted on an objective glass. This facilitates the subsequent washings and incubations.
[b] PPD mountant should be stored in the dark at –20°C until use.
[c] Incubate in a water-saturated atmosphere. This can be done by placing a hood over the cover-glasses with the inner surface coated with wet filter paper.

4.2 Surface determinants

In addition to vWF, a number of surface determinants can be used to characterize (and to purify) endothelial cells. They can all be detected by indirect immunofluorescent staining. *Table 1* lists several of these determinants as well as indicating their cellular location and whether they are present in other cell types. The antigens are detectable on living cells, however immunofluorescence studies are usually performed after fixation of the cells with 4% paraformaldehyde, methanol, or 80% acetone. The fixative used depends on the characteristics and stability of the antibodies (which are preferably monoclonal in origin).

The procedure for staining with endothelial cell-specific monoclonal antibodies is comparable with that given in *Protocol 7* for vWF, using the monoclonal antibody as the first antibody (usually incubated for 30 min) and an FITC-labelled anti-mouse antibody as the second antibody (see *Figure 2B*). A negative control using only the second antibody is necessary. *Figures 3C* and *3D* show the staining of human endothelial cells by anti-CD-31 (PECAM-1) and anti-V,E-cadherin antibodies, respectively (23, 24); these antibodies preferentially stain cell junctions. They react with large vessel

Table 1. Specific determinants of human endothelial cells[a]

Determinant	Localization in EC	Present in:
von Willebrand factor (vWF)	Intracellular granules	All EC, platelets, megakaryocytes
CD-31 (PECAM-1)	Junctions	All EC, platelets, PMN
V,E-cadherin	Junctions	All EC, endothelial macrophages
PAL-E antigen	Surface	Microvascular and venular EC
EN-4 antigen	Surface	EC
Ulex europaeus lectin-1	Surface	EC, epithelial cells
Dil-acetylated LDL uptake	Surface/lysosomes	EC and macrophages strong; other cells weak
Angiotensin converting enzyme	Surface/secreted	EC, also other cells

[a] EC, endothelial cells; PMN, polymorphonuclear leucocytes.

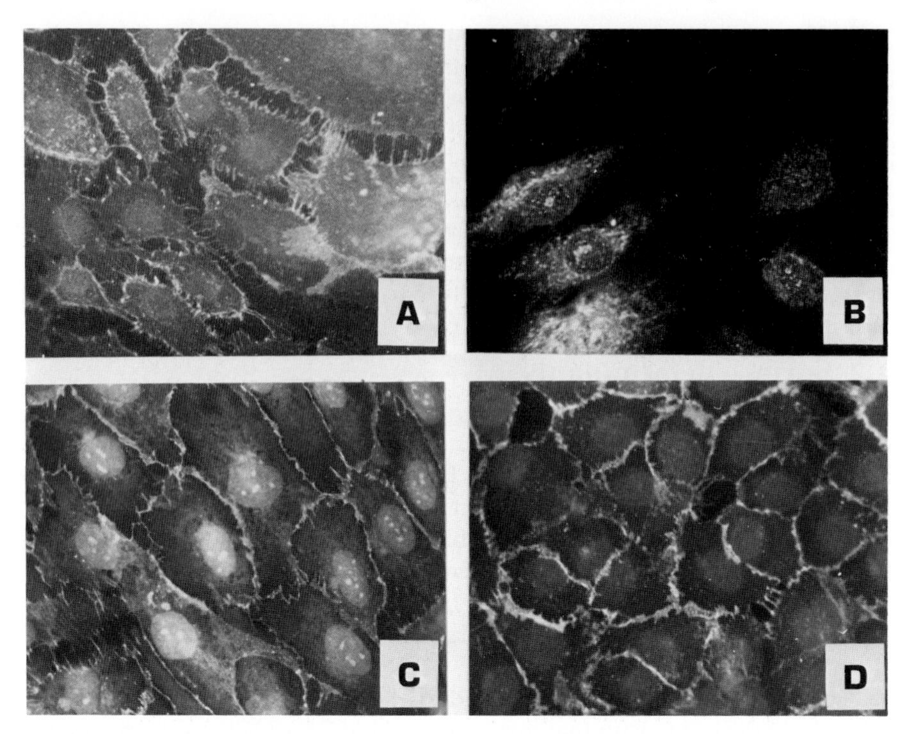

Figure 3. Characterization of human umbilical vein endothelial cells. Indirect immunofluorescent staining with the monoclonal antibodies (A) PAL-E and (B) EN-4, and showing the presence of (C) V,E-cadherin and (D) CD-31 (PECAM-1).

endothelial cells as well as with microvascular endothelial cells. *Figure 3A* displays cells stained with a monoclonal antibody PAL-E, which binds to capillary and venular endothelial cells, but not to those derived from arteries (25). As endothelial cells have different specific properties in different vascular beds and tissues, more monoclonal antibodies are becoming, and will continue to become available, that can be used to establish vessel type or tissue specificity.

Endothelial cells bind to *Ulex europaeus* lectin I which recognizes fucose-residues (26). This characteristic is shared with epithelial cells but is not present in fibroblasts and smooth muscle cells. Therefore, *Ulex europaeus* lectin I-coated Dynabeads can sometimes be used to purify endothelial cells. Direct immunofluorescent staining of endothelial cells can be performed by incubating washed (fixed) cells for 30 min at room temperature in rhodamine-labelled *Ulex europaeus* lectin I (1 mg/ml, diluted 1:10 in PBS containing $CaCl_2$ and 1 mM $MgCl_2$). After two subsequent washes with PBS, the cells are mounted with PPD mountant under a glass coverslip and observed by fluorescence microscopy (excitation 530–560 nm; emission > 580 nm).

4.3 Enzymatic activities and receptors

Another characteristic that can be used to ascertain the endothelial nature of cultured cells is the presence of specific enzymatic activities, e.g. angiotensin-converting enzyme activity (also detectable by fluorescence microscopy). In particular, when specific endothelial cell types are used, for example glomerular or brain endothelial cells, specific enzymatic activities (e.g. γ-glutamyl transpeptidase and monoamine oxidase activities in brain endothelial cells) are used, not only to characterize the cells, but also to determine whether they still retain their specific differentiated properties.

Certain receptors have also been used to provide additional evidence for the endothelial nature of cultured cells. Assay of the uptake of Dil-acetylated LDL is based on the presence of the Ac-LDL-recognizing scavenger receptor (6). Uptake of Dil-acetylated LDL (see *Protocol 6*) reflects interaction of the fluorescent Dil-acetylated LDL with the receptor, and the accumulation of the fluorescent material into multivesicular bodies/lysosomes after uptake of the Dil-acetylated LDL. It is easily detectable by fluorescence microscopy (excitation 580 nm). Dil-acetylated LDL uptake can thus be used to characterize endothelial cells in addition to other criteria. However, one has to realize that many cell types can ingest small amounts of Dil-acetylated LDL.

4.4 Criteria to determine endothelial cell activation

Endothelial cells become activated after exposure to endotoxin (LPS, bacterial lipopolysaccharide) or the cytokines, tumour necrosis factor, and interleukin-1. As a consequence they start to produce a number of new proteins, which are involved in inflammation, but are not present in quiescent endothelial cells. In addition to establishing the endothelial nature of endothelial cells, it is, therefore, valuable to ascertain that endothelial cells have not become activated in culture. Good markers for endothelial activation are E-selectin (27), a leucocyte adhesion molecule, interleukin-6 (28, 29), and urokinase-type plasminogen activator (u-PA) (30). E-selectin can be assayed by cell-ELISA in which the cell monolayer presents the antigen, and a monoclonal anti-E-selectin antibody coupled to peroxidase is used as a tagging antibody. Interleukin-6 and u-PA are secreted by endothelial cells and can be assayed in endothelial conditioned medium (stored frozen) by ELISA. While E-selectin is exclusively, but transiently, expressed on activated endothelial cells, IL-6 and u-PA can also be produced by other cell types.

5. Additional culture techniques

5.1 Storage of endothelial cells in liquid nitrogen

Human endothelial cells from various types of arteries and veins, and from foreskin microvessels, can be stored in liquid nitrogen for propagation at a later time. A simple procedure for freezing and thawing endothelial cells is given in *Protocol 8*.

Protocol 8. Storage of endothelial cells in liquid nitrogen

Equipment and reagents

- Liquid nitrogen filled container for storage of cells
- Centrifuge
- Waterbath at 37 °C
- Cryopreservator or a freezer (–80 °C) and a Tempex box (wall thickness about 2 cm)[a]
- Round-bottom plastic vials (with screw caps) for freezing cells (Greiner, cat. no. 121.261)
- Endothelial cells: confluent, early passage culture (25–75 cm^2)
- Solution A (see *Protocol 4*)
- Trypsin/EDTA (see *Protocol 4*)
- M199 medium/20% serum: 20% (v/v) human serum or NBCS (heat-inactivated) in M199 medium supplemented with 20 mM Hepes, 100 IU/ml penicillin and 100 µg/ml streptomycin (pH 7.4)
- 20% DMSO: 20% (v/v) DMSO in M199 medium/20% serum
- Culture medium (see *Protocol 4*)
- Fibronectin- or gelatin-coated dishes (see *Protocol 2*)

A. *Storage of endothelial cells in liquid nitrogen*

1. Wash the cultured cells with solution A.
2. Detach the cells by incubating with trypsin/EDTA for several minutes.
3. As soon as the cells detach, stop the trypsin activity by adding a 5-fold volume of M199 medium/20% serum.
4. Spin the cells down by centrifugating for 5 min at 200 *g*.
5. Resuspend the cells in 0.5 ml of M199 medium/20% serum.
6. Cool the cell suspension towards 0 °C (e.g. by placing it on ice).
7. Dilute the cell suspension with 0.5 ml of 20% DMSO, i.e. the final concentration of DMSO is 10% (v/v).
8. Transfer the cell suspension to a plastic round-bottom tube and screw on the cap.
9. Freeze the tube slowly (30 °C/h) to –70 °C[a].
10. Transfer the tube into liquid nitrogen, and put it in the liquid nitrogen containing cell storage container.

B. *Thawing endothelial cells*

1. Take the tubes out of the liquid nitrogen storage container and rapidly thaw the cells in a waterbath at 37 °C.
2. Dilute the cells by adding a 9-fold volume of culture medium.
3. Spin the cells down by centrifugating for 5 min at 200 *g*.
4. Resuspend the cells in culture medium and seed them on fibronectin- or gelatin-coated dishes.

5. Put the dishes into a CO_2 incubator and culture the cells as directed in *Protocols 4* and *5*.

[a]Cryopreservators are designed to carefully control the cooling of cells. However, the slow freezing of the cells to – 70°C at a rate of 1°C/2 min can be mimicked by putting the ice-cold tube containing the cells into a precooled (0–4°C) Tempex box, and freezing this box in a – 80°C freezer for 2–3 h.

5.2 Culture of human endothelial cells on porous filters

Many aspects of the regulation of endothelial cell protein synthesis can be studied in normal culture dishes. However, dish cultures are not suitable for studying the barrier function of endothelial cells (both regarding leucocyte transmigration and the exchange of fluid and macromolecules) or the polarity of protein secretion. In order to investigate these properties, endothelial cells can be grown on porous filters in an adapted Boyden chamber (see *Figure 4A*). In *Protocol 9*, a method for seeding and culturing human endothelial cells on porous filters is described. *Protocol 10* details a procedure by which the barrier function of the endothelial cell monolayer can be evaluated by measuring the passage of a marker protein (peroxidase) through it (see *Figure 4B*). Further information regarding the measurement of the passage of compounds across cell monolayers is given in Chapter 6.

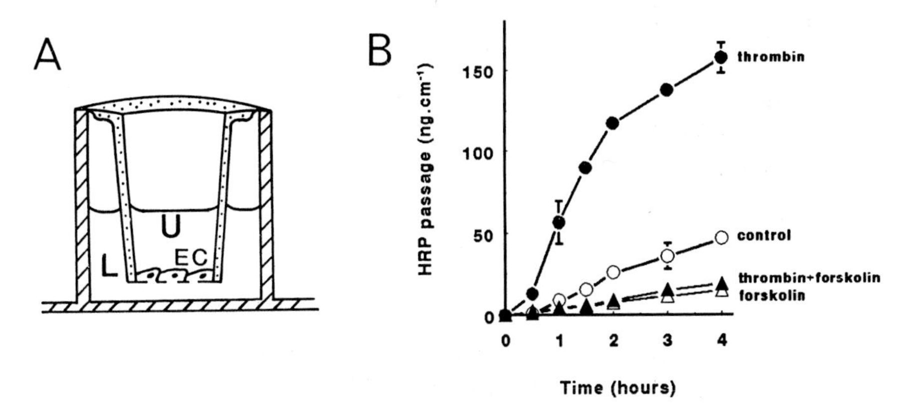

Figure 4. Assaying the permeability of human endothelial cell monolayers *in vitro*. (A) The Transwell system: endothelial cells (EC) form a tight monolayer on a porous filter (U, upper compartment; L, lower compartment). (B) Passage of horseradish peroxidase (HRP) through an endothelial cell monolayer after stimulation with 1 U/ml thrombin and/or (pre)incubation with 10 μM forskolin.

›col 9. Culture of human endothelial cells on porous filters according to the method of Langeler *et al.* (31, 32)

Equipment and reagents

- Transwell system: Transwell polycarbonate filters (pore size 3 μm; filter area 0.33 cm^2) in multiwell dishes (Costar; see *Figure 4A*)
- CO_2 incubator (water-conditioned)
- Pasteur pipette (sterile)
- Solution A (see *Protocol 4*)
- Confluent primary culture of human umbilical vein endothelial cells.
- Trypsin/EDTA (see *Protocol 4*)
- Culture medium (see *Protocol 4*)
- Fibronectin (10 μg/ml; see *Protocol 2*)

Method

1. Coat the Transwell filters with 10 μg/ml fibronectin. Place a small amount of medium under the filter so that the fibronectin solution is sucked through.
2. Aspirate the fibronectin solution from the filter immediately before seeding the cells.
3. Wash the cells with solution A and detach them by trypsin/EDTA treatment (see *Protocol 4C*).
4. Mix the suspension of detached cells with a 3-fold volume of culture medium.
5. Seed the cells on to the filters at a density that is equal to that of the confluent primary culture. Seed the cells only into the upper compartment of the Transwell system (see *Figure 4A*).[a]
6. Put the cells into an incubator at 37 °C with an atmosphere of 5% CO_2/95% air.
7. Aspirate the non-attached cells 4 h after seeding and add 1.5 ml of culture medium to the adhered cells.[b]
8. Renew the culture medium every other day.
9. Use the cells 4–7 days after seeding for experiments (e.g. *Protocol 10*).

[a]Add the suspended cells, if necessary in small volumes, until enough cells are added. If too much medium is added at once, cells may float into the lower compartment and colonize the bottom of the dish.
[b]Always add the medium to the upper compartment until it is full before adding the residual medium to the lower compartment. This avoids hydrostatic pressure underneath the cells.

5.3 Culture of human endothelial cells on microcarriers

Human endothelial cells can be cultured on Biosilon (Nunc) or Cytodex 3 (Pharmacia) microcarriers. Medium passing through a column containing endothelial cell-covered microcarriers is exposed to a large surface area of endothelial cells, particularly when solid Biosilon beads are used. This enables rapid and sensitive measurements of parameters which may normally remain below detection limit.

Protocol 10. Determination of the permeability of human endothelial cell monolayers on porous filters (32, 33)

Equipment and reagents

- CO_2 incubator (water-conditioned)
- 96-well culture dish
- Pipettes to sample 50, 100, 150, and 600 μl quantities (e.g. Pipetman P200 and P1000)
- Sterile pipette tips
- Endothelial cells grown on 12 Transwell filters (see *Protocol 9*)
- Thrombin solution: 100 U thrombin/ml PBSA
- Forskolin solution: 10 mM forskolin in 96% ethanol
- M199 medium/1% HSA with 10 μM forskolin: dilute 10 mM forskolin 1:1000 in M199 medium/1% HSA
- Pyrogen-free human serum albumin (HSA): 20% (w/v) stock solution purchased from the Central Laboratory of the Blood Transfusion Service (Amsterdam, The Netherlands)
- M199 medium/1% HSA: M199 medium (supplemented with 20 mM Hepes) with 1% pyrogen-free human serum albumin, 100 IU/ml penicillin, and 100 μg/ml streptomycin
- HRP: 5 mg/ml horseradish peroxidase in M199 medium/1% HSA
- Reagents to assay peroxidase activity (standard ELISA reagents)

Method

1. Number the filters 1–12.
2. Aspirate the culture medium from the lower and upper compartments and replace it for filters 1–6 with M199 medium/1% HSA, and for filters 7–12 with M199 medium/1% HSA with 10 μM forskolin (0.15 ml in the upper compartment and 0.6 ml in the lower compartment). Immediately replace the Transwell system into the incubator at 37 °C.
3. Prepare in M199 medium/1% HSA: 0.6 ml 150 μg/ml HRP (for filters 1–3), 0.6 ml 150 μg/ml HRP with 1 U/ml thrombin (for filters 4–6), 0.6 ml 150 μg/ml HRP with 10 μM forskolin (for filters 7–9), and 0.6 ml 150 μg/ml HRP with 1 U/ml thrombin and 10 μM forskolin (for filters 10–12). Keep 6.5 ml M199/1% HSA and 6.5 ml M199/1% HSA with 10 μM forskolin for volume replenishment of filters 1–6 and filters 7–12, respectively.
4. Take the cells out of the incubator. Start the incubation 60 min after aspiration of the medium from the upper compartments and replacement with 150 μl of the four HRP-containing solutions as indicated in step 3. Put the cells back into the incubator at 37 °C.
5. At time points 0.5, 1, 1.5, 2, 3, and 4 h: take 100 μl aliquots from the lower compartments and transfer the samples into a 96-well dish.[a] Replace the missing medium by adding (to the lower compartments) 100 μl of M199/1% HSA (filters 1–6) or M199/1% HSA with 10 μM forskolin (filters 7–12). Immediately after each sampling, replace the cells back into the incubator. Cover the samples in the 96-well dish with Parafilm.

Protocol 10. *Continued*

6. After 4 h take 50 μl samples from the upper compartments. These samples can be used to determine the HRP which was not transported through the filters and, in combination with the other samples, the total amount of HRP added to each Transwell.
7. Store the samples frozen at −20 °C until required for HRP assaying.[a]

[a]Assays for HRP can be performed using various techniques (see Chapter 3, *Protocol 4*). In calculating the passage of peroxidase through the cell monolayer, a correction has to be made for the dilution of the medium in the lower compartment by the compensatory medium replenishment (results: see *Figure 4B*).

To prepare endothelial cell-covered beads, the microcarriers are coated with human fibronectin (10 μg/ml; see *Protocol 2*). Endothelial cells, which have been detached from dishes by treatment with trypsin/EDTA, are suspended in culture medium and mixed with the microcarriers at an appropriate concentration. The mixture is incubated in a tube at 37 °C under a 5% CO_2/95% air atmosphere, and gently moved at 10 min intervals. After 2 h, the microcarriers are poured into a bacteriological Petri dish and maintained under standard culture conditions (see *Protocol 4*) until required for experimental purposes.

Microcarriers can be transferred into a Techne 500 ml microcarrier stirrer flask (Techne, cat. no. 7609). The microcarriers are gently suspended by intermittent stirring (1 min stirring in every 4 min) at 25 r.p.m. (Techne stirrer MCS-104L and Techne waterbath MWB 10L). The whole system is kept under a 5% CO_2/95% air atmosphere (supplied from a gas cylinder via a water flask, a 0.22 μm filter, and sterile tubes). One-half of the culture medium is renewed daily. Stirring must be done as gently as possible because human endothelial cells are easily damaged.

6. Perspective

Endothelial cells from large blood vessels can nowadays be cultured on a routine basis. It is also possible to culture endothelial cells from a number of microvascular beds. These cell cultures provide a powerful tool to study the properties of endothelial cells in detail by biochemical means without the interference of other tissue cells, and have contributed significantly to our present insight into the role of endothelial cells in vasoregulation, haemostasis, leucocyte adhesion and extravasation, barrier function and angiogenesis. It is important to appreciate, however, that certain differentiated properties may be lost during the isolation and culture of these cells. Co-cultures between endothelial cells and pericytes, astrocytes, or smooth muscle cells may partly overcome this loss, but this has still to be established. Although the final evaluation of the importance of *in vitro* data can only be derived from *in/ex vivo* studies, the flexibility of these *in vitro* systems will help to unravel the molecular properties and metabolic regulation of the pluripotent endothelial cell.

Acknowledgement

Richard Draijer was supported by the Netherlands Heart Foundation, Grant no. 90.085.

References

1. Jaffe, E. A., Nachman, R. L., Becker, C. G., and Minick, C. R. (1973). *J. Clin. Invest.*, **52**, 2745.
2. Gimbrone, M. A. Jr., Cotran, R. S., and Folkman, J. (1974). *J. Cell Biol.*, **60**, 673.
3. Maciag, T., Cerundolo, J., Ilsley S., Kelley, P. R., and Forand, R. (1979). *Proc. Natl Acad. Sci. USA*, **76**, 5674.
4. Burgess, W. H., Mehlman, T., Friesel, R., Johnson, W. V., and Maciag, T. (1985). *J. Biol. Chem.*, **260**, 11389.
5. Vuento, M. and Vaheri, A. (1979). *Biochem. J.*, **183**, 331.
6. Basu, S. K., Goldstein, J. L., Anderson, R. G. W., and Brown, M. S. (1976). *Proc. Natl Acad. Sci. USA*, **73**, 3178.
7. Redgrave, T. G., Roberts, D. C. K., and West, C. E. (1974). *Anal. Biochem.*, **65**, 42.
8. Van Hinsbergh, V. W. M., Scheffer, M. A., and Langeler, E. G. (1990) In *Cell culture techniques in heart and vessel research* (ed. H. M. Piper), pp. 178–204. Springer–Verlag, Berlin.
9. Thornton, S. C., Mueller, S. N., and Levine, E. M. (1983). *Science*, **222**, 623.
10. Maciag, T., Hoover, G. A., Stemerman, M. B., and Weinstein, R. (1981). *J. Cell Biol.*, **91**, 420.
11. Van Hinsbergh, V. W. M., Mommaas-Kienhuis, A. M., Weinstein, R., and Maciag, T. (1986). *Eur. J. Cell. Biol.*, **42**, 101.
12. Hoshi, H. and McKeehan, W. L. (1986). *In Vitro Cell. Dev. Biol.*, **22**, 51.
13. Van Hinsbergh, V. W. M., Binnema, D., Scheffer, M. A., Sprengers, E. D., Kooistra, T. and Rijken, D. C. (1987). *Arteriosclerosis*, **7**, 389.
14. Davison, P. M., Bensch, K., and Karasek, M. A. (1980). *J. Invest. Dermatol.*, **75**, 316.
15. Voyta, J. C., Via, D. P., Butterfield, C. E., and Zetter, B. R. (1984). *J. Cell Biol.*, **99**, 2034.
16. Davison, P. M., Bensch, K., and Karasek, M. A. (1983). *In Vitro*, **19**, 937.
17. Carley, W. W., Niedbala, M. J., and Gerritsen, M. E. (1992). *Am. J. Respir. Cell Mol. Biol.*, **7**, 620.
18. Gerhart, D. Z., Broderius, M. A., and Drewes, L. R. (1988). *Brain Res. Bull.*, **21**, 785.
19. McGee-Russel, S. M. and De Bruijn, W. C. (1971). In *Cell structure and its interpretation* (ed. S. M. McGee-Russel and W. C. De Bruijn), pp. 115–33. Arnold, London.

20. Weibel, E. R. and Palade, G. E. (1964). *J. Cell Biol.*, **23**, 101.
21. McEver, R. P., Beckstead, J. H., Moore, K. L., Marshall-Carlson, L., and Bainton, D. F. (1989). *J. Clin. Invest.*, **84**, 92.
22. Bonfanti, R., Furie, B. C., Furie, B., and Wagner, D. D. (1989). *Blood*, **73**, 1109.
23. Albelda, S. M., Muller, W. A., Buck, C. A., and Newman, P. J. (1991). *J. Cell Biol.*, **114**, 1059.
24. Lampugnani, M. G., Resnati, M., Raiteri, M., Pigott, R., Pisacane, A., Houen, G., Ruco, L. P., and Dejana, E. (1992). *J. Cell Biol.*, **118**, 1511.
25. Schlingemann, R. O., Dingjan, G. M., Emeis, J. J., Blok, J., Warnaar, S. O., and Ruiter, D. J. (1985). *Lab. Invest.*, **52**, 71.
26. Holthöfer, H., Virtanen, I., Kariniemi, A. -L., Hormia, M., Linder, E., and Miettinen, A. (1982). *Lab. Invest.*, **47**, 60.
27. Bevilacqua, M. P., Stengelin, S., Gimbrone, M. A., and Seed, B. (1989). *Science*, **243**, 1160.
28. Sironi, M., Breviario, F., Proserpio, P., Biondi, A., Vecchi, A., Van Damme, J., Dejana, E., and Mantovani, A. (1989). *J. Immunol.*, **142**, 549.
29. Jirik, F. R., Podor, T. J., Hirano, T., Kishimoto, T., Loskutoff, D. J., Carson, D. A., and Lotz, M. (1989). *J. Immunol.*, **142**, 144.
30. Van Hinsbergh, V. W. M., Van den Berg, E. A., Fiers, W., and Dooijewaard, G. (1990). *Blood*, **75**, 1991.
31. Langeler, E. G. and Van Hinsbergh, V. W. M. (1988). *Thromb. Haemostas.*, **60**, 240.
32. Langeler, E. G., Snelting-Havinga, I., and Van Hinsbergh, V. W. M. (1989). *Arteriosclerosis*, **9**, 550.
33. Langeler, E. G. and Van Hinsbergh, V. W. M. (1991). *Am. J. Physiol.*, **260**, C1052

6

Studying transport processes in absorptive epithelia

PER ARTURSSON, JOHAN KARLSSON, GÖRAN OCKLIND, and NICOLAAS SCHIPPER

1. Introduction

Transport processes in epithelia have traditionally been studied in a variety of organ and tissue preparations from different species. However, native epithelial tissues are complex and contain many different cell types. Monolayer cultures from established epithelial cell lines have the advantages of being structurally simple, homogeneous, and easy to manipulate. Transport studies in epithelial monolayers are, therefore, often easier to perform and interpret (1). The major disadvantage of epithelial cell cultures is that in most situations they are unlikely to serve as a complete substitute for the more complex normal epithelium. Moreover, epithelial cells in culture may express carriers and enzymes in a different or more variable manner than seen *in vivo*. Despite these disadvantages, recent studies in epithelial cell monolayers have brought new insights into the functions of a variety of transport mechanisms of endogenous as well as exogenous molecules such as drugs. The aim of this chapter is to present some of the protocols used in our laboratory to study passive and active drug transport processes across monolayers of the human intestinal epithelial cell line Caco-2 grown on permeable supports (Figure 1). However, before such studies are performed it is important to know how to culture and assess the integrity of the cell monolayers on permeable supports in a reproducible manner. Methods for such characterization are presented initially. The methods for studying integrity and transport processes are general and should, therefore, be applicable to solutes other than drugs and to epithelial cell cultures other than Caco-2.

Caco-2 cells, which are derived from a human colonic adenocarcinoma, exhibit morphological as well as functional similarities to intestinal (absorptive) enterocytes (2). The cells form tight junctions and express many brush border enzymes, some CYP isoenzymes, and phase II enzymes such as glutathione-*S*-transferases, sulfotransferase, and glucuronidase, i.e. enzymes relevant in studies of presystemic (drug) metabolism (3–8). Many active

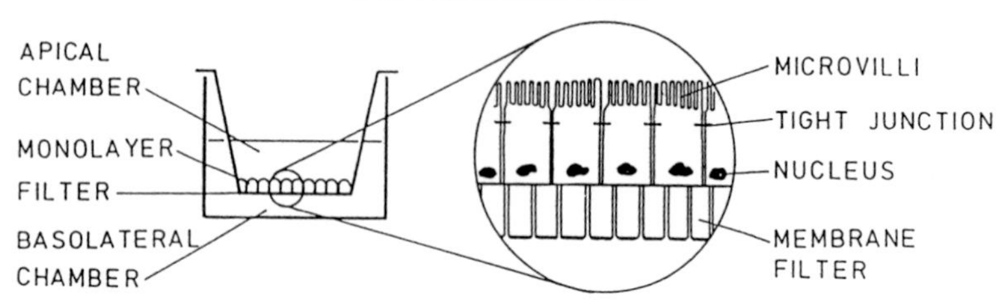

Figure 1. Diagram of an epithelial cell monolayer grown on a permeable support in a cell culture insert.

transport systems that are normally found in small intestinal enterocytes have been characterized. These include transport systems for amino acids, dipeptides, vitamins, and cytostatics (9–12). The unusually high degree of differentiation, together with the fact that these cells differentiate spontaneously in normal serum-containing cell culture medium, has resulted in Caco-2 becoming one of the most popular cell lines in studies of epithelial integrity and transport.

2. Culture of Caco-2 cell monolayers

2.1 Sources of cells

Most investigators have used the parent polyclonal Caco-2 cell line obtainable from the American Type Culture Collection (ATCC). Recently, an increasing number of investigators have cloned the parent cell line in order to obtain cell populations with more specific properties such as a higher expression of carriers for bile acids or dipeptides (13, 14). The clones are normally not available through general sources but can often be obtained from the individual investigators.

2.2 Growth and maintenance of Caco-2 cells

Caco-2 cells are generally grown in serum-containing cell culture medium (see *Protocol 1*) although a serum-free medium has been developed (15). The cells are grown and expanded in normal tissue culture flasks. A special feature of Caco-2 cells is that they attach strongly to the plastic of the tissue culture flasks. Therefore, vigorous trypsinization is needed in order to remove all cells from the plastic. We have noted that adherence to the flasks is less in subconfluent cultures. Therefore, we trypsinize the cells at 80% confluence. Complete removal of the cells from the tissue flasks is especially important when the parent Caco-2 cell line is used, otherwise less adherent cell clones may be selected in each passage resulting in an altered phenotype. Even if the cells are completely removed from the flasks during each passage, some clones may grow more rapidly than others, resulting in the selection of rapidly growing clones. Because of this, the Caco-2 cells should be used within a limited number of passages. We generally use Caco-2 cells between passages 90 and 105. In order to detect contamination, the medium is not supplemented with antibiotics in the maintenance cultures. In addition, we routinely screen our cell cultures for mycoplasma contamination every second month.

A variety of permeable supports for epithelial cell cultures are available from different suppliers (see Chapter 3, *Table 2*). We prefer to use supports made of polycarbonate since:

(a) Caco-2 cells form tight and well-differentiated monolayers on these filters without the need for extracellular matrix proteins such as collagen (16).
(b) This material is relatively hydrophilic, which generally limits the adsorption of solutes to the support which could complicate the interpretation of transport experiments (17).
(c) The polycarbonate support does not restrict the diffusion of water, as has recently been reported for some other permeable supports (18).

The pore diameter (0.4 μm) is chosen to be small enough to avoid transmigration of Caco-2 cells from the apical to the basolateral side. Penetration through permeable supports with a mean pore diameter of 3.0 μm and subsequent cell growth of a second monolayer on the basolateral side of the support has been reported for Caco-2 cells (19). Since Caco-2 cells are grown for much longer on the permeable supports than in the tissue culture flasks, we supplement the medium with antibiotics in order to avoid contamination.

Protocol 1. Culture of Caco-2 cells

Equipment and reagents

- Caco-2 cells (ATCC)
- For culture in tissue culture flasks: Dulbecco's modified Eagle's medium (DMEM; 4.5 g/litre glucose; pH 7.4) supplemented with 1% (v/v) non-essential amino acids and 10% fetal calf serum (FCS; all media from Gibco–BRL)
- Trypsin/EDTA solution (pH 6.4): 0.2% EDTA and 0.25% trypsin in PBS without Ca^{2+} and Mg^{2+} (PBSA) (Gibco) (Chapter 3, *Protocol 1*)
- 75 cm^2 tissue culture flasks (Costar or Nunc)
- Transwell polycarbonate cell culture inserts, diameter 12 mm, pore diameter 0.4 μm (Costar)
- Tissue culture plates containing 12 wells (Costar)
- For culture on permeable supports: Dulbecco's modified Eagle's medium (DMEM; 4.5 g/litre glucose; pH 7.4) supplemented with 1% (v/v) non-essential amino acids, 10% fetal calf serum (FCS), benzylpenicillin (100 IU/ml) and streptomycin (100 μg/ml; all media from Gibco–BRL)

A. *Maintenance culture in tissue culture flasks*

1. Maintain the cells in tissue culture flasks until they reach 80% confluence. Use 30 ml complete medium in each 75 cm^2 tissue culture flask and change the medium every second day. The cells are cultured at 37 °C and 10% CO_2 in 95% relative humidity.

Protocol 1. *Continued*

2. Subculture the cells as follows.
   (a) Remove the medium, wash the cells with 15 ml PBSA and add 5 ml of the trypsin solution.
   (b) Wash all inner walls of the flask and remove all but 1 ml of the trypsin solution.
   (c) Detach the cells at 37 °C (5–10 min).
   (d) Stop the trypsinization by adding 10 ml of complete medium.
   (e) Remove one drop for cell counting and viability testing with Trypan Blue. Centrifuge the cells for 5 min at 150 *g*. Meanwhile, count and determine the viability of the cells (> 95% viability is required). An 80% confluent 75 cm^2 flask gives approximately $1–2 \times 10^7$ cells.
   (f) Dilute 1:10 for subculture in new flasks or to 1×10^6 cells/ml for subculture on to permeable supports. Use medium supplemented with antibiotics if the cells are to be seeded on permeable supports.

B. *Culture on permeable cell culture inserts*

1. Place the desired number of Transwells in the wells of the tissue culture plates and add 0.1 ml of complete medium to the apical and 1.5 ml to the basolateral sides of the Transwell. Wet the filters with medium for 30 min in the CO_2 incubator. This step can be performed before the harvest of the cells from the flasks.
2. Resuspend the cell suspension (1×10^6 cells/ml, see step 2f above) and add 0.5 ml of cells to the apical side of the Transwell. Avoid spilling cells in the basolateral chamber.
3. Keep the cells in the CO_2 incubator for 3 h. Change the apical medium (0.6 ml).
4. Change the medium every second day for 21–30 days to obtain well-differentiated monolayers suitable for transport experiments.

2.3 Assessment of monolayer integrity

2.3.1 Transepithelial electrical resistance (TER)

Qualitative assessment of epithelial integrity can be performed by measuring TER. This is most conveniently achieved directly in the cell culture inserts with specially designed electrodes (e.g. the Millicell ERS from Millipore or the EVOM from WPI). The sensitivity of these electrodes is limited and they should, therefore, not be used for quantitative determination of electrical resistance. For instance, we have found that the resistance is extremely sensitive to the positioning of the electrodes. However, there are several reports of the use of these electrodes for the assessment of large relative changes in TER with, for example, time. More sophisticated equipment is needed for quantitative studies of changes in resistance. For a more complete discussion on electrophysiological assessment of epithelia, see a previous volume in the *Practical approach series* (20).

The electrical resistance across Caco-2 monolayers should increase with time, reaching a maximum after about 10 days in culture (17). Resistances ranging from 150 ohm/cm^2 to up to 600 ohm/cm^2 have been reported. These differences may be related to a lack of temperature control, the use of different clones or passages of Caco-2 cells, or differences in the cell culture conditions. In our laboratory, variability in resistance values between individual monolayers of the same batch is typically <20%.

2.3.2 Transport of radiolabelled mannitol

An alternative method of assessing monolayer integrity is to use a radiolabelled hydrophilic marker molecule which is passively transported across the monolayers (see *Protocol 2*). Passive transport of molecules across Caco-2 monolayers can either occur across the cell membranes (the transcellular route), or through the tight junctions bridging the intercellular spaces between the cells (the paracellular route) (*Figure 2*). Since hydrophilic molecules are not distributed to the cell membrane to a large extent, their transport across the epithelial monolayers is limited to the paracellular route. Thus, the permeability of a hydrophilic marker molecule should decrease with time in culture, reaching a minimum in confluent monolayers with fully developed tight junctions.

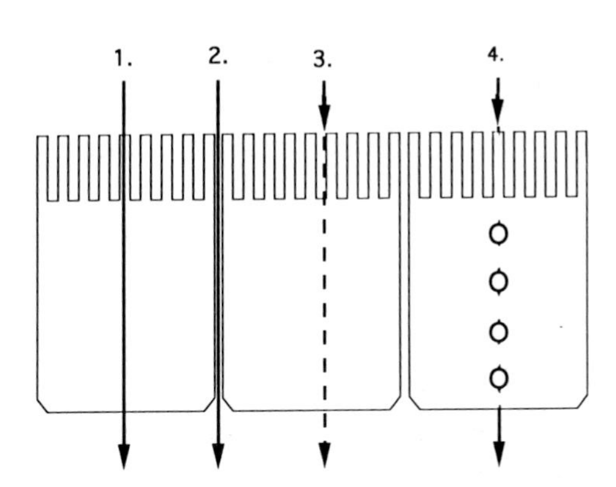

1. Passive transcellular
2. Passive paracellular
3. Active transcellular (carrier-mediated)
4. Transcytosis

Figure 2. Transport routes in epithelia. Routes 1–3 are dealt with in this review.

The tight junction barrier displays both size and charge selectivity. The permeability of tight junctions decreases with the molecular weight of the marker and is higher for positively charged molecules. To increase the sensitivity, and to avoid variability in charge selectivity we have chosen a relatively small and uncharged molecule, mannitol (M_w = 182 g/mole), as a marker of epithelial integrity. In our experience, ^{14}C-labelled mannitol has a several-fold lower permeability than ^{3}H-labelled mannitol, even after evaporation of tritiated water from the latter (21). The transport of [^{14}C]mannitol across mature intact Caco-2 monolayers should be $< 0.05\%$/h. We generally calculate the apparent permeability coefficient (P_{app}) of [^{14}C]mannitol as a measure of the integrity of the Caco-2 cell monolayers. This parameter has the advantage of being independent of experimental design, surface area, time of experiment, and driving concentration, which makes it possible to compare P_{app} values obtained under quite different experimental situations. We have, for instance, made direct comparisons of P_{app} values from Caco-2 monolayers, rat intestinal segments, and the perfused human intestine (22). One drawback of using the permeability coefficient is that there may be limited information about the surface area of complex epithelial tissues.

As for measurements of transepithelial resistance, it is very important to control temperature and evaporation of solutions during the experiment. All media should therefore be prewarmed in a waterbath and the monolayer should be kept in a humid box at 37°C. The experiments are performed in air. In order to avoid rapid cooling during sampling, we place the tissue culture plates on thermostatically controlled plate warmers. The lowest temperature variation between the wells of the culture plate is obtained using the Microtemp plate warmer/cooler from Detrona, especially designed for cell culture plates.

Protocol 2. Assessment of epithelial integrity with the hydrophilic marker mannitol

Equipment and reagents

- [^{14}C]mannitol, specific activity: 55 mCi/mmol; 0.1 mCi/ml; 1.8 µmol/ml (New England Nuclear)
- HBSS; Hanks' balanced salt solution (pH 7.4; Gibco–BRL) supplemented with 25 mM Hepes (Sigma)
- Caco-2 cells cultured in Transwell inserts for 21–30 days as described in *Protocol 1*
- Microtemp plate heater (Detrona)

Method

1. Remove the complete cell culture medium and pre-incubate four cell monolayers with prewarmed HBSS for 30 min. This mild washing step adapts the monolayers to the new buffer and eliminates residual enzyme activity from serum. Residual esterase activity may cause hydrolysis of prodrugs prior to transport.

2. Add trace amounts of [^{14}C]mannitol to prewarmed HBSS (corresponding to a final amount of approximately 1×10^6 c.p.m./cell culture insert).
3. Add 0.4 ml of [^{14}C]mannitol containing HBSS to the apical side and 1.5 ml of HBSS to the basolateral side of the monolayers. Take 10–50 μl samples from the apical chambers at $t = 0$ for determination of the initial radioactivity of each filter.
4. Transfer the inserts to new wells containing fresh HBSS at 30 min intervals for 150 min. The choice of correct time interval is critical and may vary depending on the integrity of the Caco-2 monolayers.
5. At the end of the experiment, take 10–50 μl samples from the apical chambers and 0.5–1.0 ml samples from the basolateral chambers.
6. Determine the radioactivity of the samples in a liquid scintillation counter.
7. Plot the cumulative fraction absorbed, FA_{cum}, versus time. FA_{cum} is defined as:

$$FA_{cum} = \sum \frac{C_{R_i}}{C_{D_i}};$$

where C_{R_i} is the receiver concentration at the end of interval i and C_{D_i} is the donor concentration at the beginning of interval i.

A linear relationship should be obtained. Use the slope or transport rate constant, k (min^{-1}), to calculate the apparent permeability, P_{app} (cm/s):

$$P_{app} = \frac{k \cdot V_R}{A \cdot 60},$$

where V_R is the volume (ml) of the receiver chamber and A is the surface area (cm^2) of the cell culture insert.[a]
8. Calculate the total amount of radioactivity, i.e. the sum of the radioactivity in the apical and basolateral chambers.[b]

[a] The P_{app} value for [^{14}C]mannitol in intact Caco-2 monolayers should generally be $< 1 \times 10^{-6}$ cm/s. However, it may be > 10 times lower, depending on such aspects as the batch of FCS, the passage number of the cells, or the culture conditions. The standard deviation (n = 4) of P_{app} is generally < 10%.

[b] The recovery of a hydrophilic marker such as mannitol is typically > 98%.

2.3.3 Visualization of epithelial integrity

Visualization of cultured cell monolayers is an important additional means of assuring their integrity. It can be used to verify the homogeneity and tightness of the cultured epithelium. With microscopic examination it is possible to distinguish whether an increased permeability is the result of loss of cells or of more general discontinuities in the monolayer. Microscopic techniques may also be used to investigate whether manipulations of the cells, such as addition of transport inhibitors and pharmaceutical additives, change the barrier properties of the monolayer (23). Unfortunately, direct visualization by light microscopy is not possible with normal uncoated Transwell cell culture inserts due to the high pore density in the filter membrane of the inserts which scatters the illuminating light. However, special inserts with a lower pore density (Transwell Clear and Falcon Cyclopore cell culture inserts) may be used. These inserts allow direct inspection of the growth and development of the monolayer with a phase-contrast microscope. The lower density of pores, however, limits the transfer of compounds to the basolateral side of the cells.

Fluorescence microscopy is a more powerful tool for examining integrity and changes in barrier properties at a cellular level. General fluorescent markers such as nucleus staining and actin binding agents can be used for many cell lines, while antibody type markers are specific only for the proteins they have been raised against. For example, tight-junction specific antibodies are often produced against Madin–Darby canine kidney (MDCK) cell proteins and have a low cross-reactivity against human adenocarcinoma cell lines (see *Protocol 5*).

The fluorescent agent, propidium iodide, stains the cell nucleus, but does not penetrate normal living cell membranes. It may, therefore, be used to discern cell viability, or to study the effects of additives on cell membrane permeability (24, 25) (see *Protocol 3*).

Protocol 3. Assessment of membrane integrity with the fluorescent marker, propidium iodide

Equipment and reagents

- Fluorescence microscope with filter for propidium iodide (excitation = 536 nm; emission = 617 nm (bound to DNA))
- Propidium iodide (Molecular Probes)
- Caco-2 cell monolayers grown on Transwell cell culture inserts for 21–30 days as described in *Protocol 1*
- Glycerol (microscopy grade, Sigma)/PBS 1:1
- PBS (pH 7.5)
- Fine-pointed scalpel
- Paraformaldehyde (PF) (3% w/v) (E. Merck) in PBS

Method

1. Rinse Caco-2 cell monolayers apically and basolaterally twice on each side with PBS.[a]
2. Incubate the monolayers apically with freshly prepared propidium iodide (15 μg/0.5 ml PBS) for exactly 3 min.
3. Rinse as in step 1.
4. Fix the cells for 10 min in PF, and rinse four times as in step 1.

5. Rinse the cells once in PBS/glycerol solution, and carefully cut out the monolayer and supporting filter membrane from the cell culture insert with a fine-pointed scalpel. Mount in glycerol/PBS under a coverglass. Examine the monolayer under the fluorescence microscope.[b]

[a] Steps 1–3 should be performed quickly but carefully since the cells must remain viable to avoid artificial uptake of the fluorescent probe.
[b] Analysis of the sample should be performed immediately after preparation of the monolayers for best results. Approximately 0.1% of the cells will be stained in normal viable cell monolayers.

Actin staining with rhodamine-labelled phalloidin has been used as an indirect method to study the integrity of tight junctions since changes in actin staining often (but not always) correlate with changes in tight-junction properties (25, 26) (see *Protocol 4*). Visualization of components more closely related to the tight junctions will, however, give more information about the integrity of the tight junctions and effects of additives on the barrier properties. Antibodies raised against the tight-junction protein ZO-1 provide markers to stain such a component (see *Protocol 5*) (27). The monoclonal rat antibody R40.76 is to date the only commercially available antibody directed against the tight-junction protein ZO-1 (see Section 4). It is developed against the mouse ZO-1 protein and was claimed to bind to ZO-1 from several species but not to the human protein. However, Zahraoui *et al.* (28) recently reported that the monoclonal antibody (mAb) R40.76 binds to Caco-2 cells fixed and permeabilized with methanol at –20°C. This method was tested in our laboratory, together with several other fixation protocols including drying. However, we could not detect ZO-1 in the Caco-2 cells, even after prolonged incubation with high titres of R40.76. Finally, the protocol presented here was developed (see *Protocol 5*). We found that the antibody has a low affinity for the native human ZO-1 protein and, therefore, a long incubation time is needed. Consequently, fixation and drying are deleterious. The same protocol may be used for cryosections provided the sections are not allowed to dry but are immediately soaked in buffer after sectioning.

One disadvantage of the above staining techniques is that they may result in specimens in which it is difficult to localize the cells or that are hard to focus under the microscope. This may be overcome by the use of the nucleus staining agent, bisbenzimide, as a second fluorescent marker. This compound passes cell membranes, and fluoresces brightly on binding to DNA at a wavelength that does not interfere with the primary markers (see *Protocol 5*). Additionally, this agent may be used to determine cell density and distribution in cultured monolayers (29).

Transmission electron microscopy is a valuable but time- and money-consuming tool in visualizing cellular morphology, especially at a subcellular level (23) (see Chapter 3, Section 4.2). It may be used for more detailed study of cell morphology, such as density of microvilli, development of tight junctions, and the status of cell organelles and cell membranes. Since normal tight junctions are impermeable to electron-dense markers such as Ruthenium Red and horseradish peroxidase, the exclusion of these markers from the paracellular spaces below the tight junction can be used to assess paracellular permeability (30, 31).

Protocol 4. Actin staining with rhodamine phalloidin in Caco-2 cell monolayers

Equipment and reagents

- Fluorescence microscope with filter for rhodamine (excitation, 550 nm; emission, 580 nm)
- Rhodamine phalloidin (Molecular Probes)
- Paraformaldehyde (PF), 3% (w/v) (Merck) in PBS
- PBS (pH 7.4)
- Triton X-100, 1% (v/v) (Sigma) in water
- Caco-2 cell monolayers grown on Transwell cell culture inserts for 21–30 days as described in *Protocol 1*
- Glycerol (microscopy grade, Sigma)/PBS 1:1
- Fine-pointed scalpel

Method

1. Rinse Caco-2 cell monolayers apically and basolaterally twice each side with PBS.
2. Fix the cells for 10 min in PF, and rinse four times as in step 1.
3. Permeabilize the cells in Triton X-100 for 10 min at 4°C, rinse twice as in step 1, and let the filters dry in air.
4. Incubate the monolayers apically with freshly prepared rhodamine phalloidin (5 µl/0.2 ml PBS) in the dark for 20 min at room temperature.
5. Rinse the cells three times with PBS as in step 1.
6. Rinse the cells once in PBS/glycerol solution, and carefully cut out the monolayer and supporting filter membrane from the cell culture insert with a fine-pointed scalpel. Mount in glycerol/PBS under a coverglass. Examine the monolayer under the fluorescence microscope.[a]

[a] Analysis of the sample should preferably be performed after preparation of the monolayer, but the actin staining is fairly stable for a few weeks at 4°C.

Protocol 5. Immunofluorescence of the tight-junction protein ZO-1 in Caco-2 monolayers

Equipment and reagents

- PBS (pH 7.5)
- Bovine serum albumin (BSA) (fraction V, globulin free, Sigma)
- Fine-pointed scalpel
- Glycerol (microscopy grade, Sigma)/PBS 1:1

- Caco-2 cell monolayers grown on 12 mm diameter Transwell filters for 21–30 days as described in *Protocol 1*. Preferably, use collagen-treated filters, otherwise cells are prone to detach during incubation with the ZO-1 antibody (step 2 below)
- Paraformaldehyde (PF) (Merck) 3% (w/v) in PBS, pH adjusted to 7.5. Use freshly made
- Streptavidin–Texas Red (Amersham, cat. no. RPN 1233)
- Biotinylated sheep anti-rat IgG (Amersham, cat. no. RPN 1002)
- Rat anti-ZO-1 mAb R40.76 (Chemicon Int.). The rat anti-ZO-1 mAb R40.76 is supplied as ascites fluid. Dilutions of 200–800 times should give a good signal
- Shaker table (Titertec, Flow Labs)
- Triton X-100 (Sigma)
- bisbenzimide (Molecular Probes, H-33258), 0.5 mg/ml in water (stock solution)
- Fluorescence microscope with filters for rhodamine/Texas Red (excitation, 550 nm; emission, 580 nm) and bisbenzimide (excitation, 346 nm; emission, 460 nm)

Method

1. Rinse cells twice in PBS at room temperature.
2. Incubate with R40.76 mAb diluted in PBS supplemented with 0.2% BSA and 0.2% Triton X-100, for 12 h at 4°C.[a]
3. Wash off excess antibody solution and wash once more for 10 min in BSA/PBS on shaker table at 60 r.p.m. Use 2 ml/well of medium distributed on both sides of the filter. Be careful not to disturb the cell layer which, before fixation, may be easy to dislodge.
4. Fix the cells in 3% PF for 10 min at room temperature. Wash as in step 3. This procedure will make the cells stick better to the substrate during the following steps.
5. Incubate with biotinylated sheep anti-rat IgG 1:250–1:500 for 45 min at room temperature (0.5 ml/filter), and wash as in step 3.
6. Incubate with streptavidin–Texas Red 1:100 plus 1 μg/ml of bisbenzimide (nuclear staining) for 30 min (0.5 ml/filter), and wash twice for 10 min each time.[b]
7. Rinse the cells once in PBS/glycerol solution, and carefully cut out the monolayer and supporting filter membrane from the cell culture insert with a fine-pointed scalpel. Mount in glycerol/PBS under a coverglass. Examine the monolayer under the fluorescence microscope.

[a] Perform the staining procedure in the inserts and use a 12-well culture plate. Add 0.5 ml of immunoreagents on the apical side of the filter inserts.
[b] The use of bisbenzimide results in bright blue fluorescence of the nuclei, making it easy to focus the specimen and to determine the density and integrity of the cell layer.

3. Transport studies

Transport processes across absorptive epithelia are mediated through one or several of the four routes described in *Figure 2*. In this section we shall discuss methods for studying passive and carrier-mediated transport processes. Carrier-mediated transport processes are usually divided into facilitated diffusion and active transport (32). Here we only describe methods for the initial characterization of active transport. Moreover, we assume that the compound is not metabolized during the transport (33). Methods for studying transcytosis across epithelial monolayers have been presented in a previous volume of this series (34). Methods for the study of active ion transport across epithelial monolayers have also been published previously (35).

Transport studies in Caco-2 monolayers are largely performed according to *Protocol 2*. Normally, the compound is added to the apical or basolateral side, the appearance of the compound on the opposite side is followed with time, and the transport rate is calculated. When passive transport is studied the transport rate is expressed as a P_{app} value. In the case of active transport it is more common to calculate flux rate constants (see *Protocol 6*). The transport studies should be performed under 'sink' conditions in order to reduce the effect of drug molecules diffusing back from the basolateral to the apical side or vice versa. Therefore, the duration of the experiment may vary from a few minutes for rapidly transported compounds (e.g. lipophilic drugs) to a few hours for slowly transported compounds (e.g. hydrophilic peptides). Calculation of the mass balance (i.e. the total yield of the compound) at the end of the transport experiment, will give information about possible evaporation of water, cellular accumulation of the solute, and indicate if the compound is lost as a result of adsorption to the plastic. By varying the experimental conditions, it is possible to discriminate between passive and active transport. Since cellular metabolism is inhibited at 4 °C, transport at this temperature is normally passive and the relative contribution of active and passive transport to the overall transport of a substance can be quantified simply by subtracting the transport at 4 °C from that at 37 °C (36).

Unfortunately, low temperatures also decrease the passive permeability of Caco-2 monolayers slightly, so complete independence from temperature is seldom observed (37). Still, the use of low temperatures is a simple and well-established method for transport inhibition. Alternatively, the active net transport can either be calculated directly as described in *Protocol 6*, or be estimated by subtracting the flux in one direction from that in the other (33). There is, as yet, no simple direct method available for the determination of the relative contribution of passive transcellular and paracellular transport.

When the permeabilities of unlabelled substances are studied, the drug samples are usually analysed by HPLC. In this case we generally inject the samples from the receiving chamber directly on to the HPLC without an extraction step. This procedure is particularly useful in the screening of homologous series of compounds, such as drug substances, where the same HPLC method can be used repeatedly. HPLC analysis will also reveal whether the substance is metabolized during transport.

3.1 Passive transport

Passive transport is driven by external driving forces such as differences in concentration or electrical potential. It typically displays non-saturable kinetics, cannot be inhibited by structural analogues, and does not require metabolic energy. We generally perform the following experiments (according to *Protocol 2*) to investigate whether the transport is passive:

(a) Determination of P_{app} in the apical to basal and basal to apical directions. The P_{app} should be largely independent of the transport direction. Note that a small difference may be observed due to differences in the surface areas of the apical and basolateral sides of the monolayer.

(b) Determination of P_{app} over a wide concentration interval (e.g. 10^{-6}–10^{-3} M). P_{app} should be independent of concentration. Note that higher concentrations of some substances may be toxic to the cells. If the toxicity of a substance is unknown, the effect of the substance on TER or [^{14}C]mannitol permeability should be investigated.

(c) Determination of P_{app} in the presence and absence of more or less specific transport inhibitors. Many investigators initially use a low temperature (4 °C) as a general inhibitor of active transport.

3.2 Active transport

In contrast to passive transport, active transport is still present in the absence of external driving forces. Active transport should display saturation kinetics, show substrate specificity, function against a concentration gradient, and be dependent on metabolic energy. It can either be driven directly by ATPases such as Na/K-ATPase (primary active transport), or by flux coupling to a second substance which undergoes primary active transport (e.g. transepithelial transport of glucose which is coupled to primary active Na^+ transport) (32).

Active transport in epithelial cell monolayers may involve distinct carriers in both the apical and basolateral cell membranes (38). The characterization of each of these carriers is more precise if performed in apical and basolateral membrane vesicles than in epithelial cell monolayers. Caco-2 cell monolayers have, therefore, mainly been used in studies of the active net transport of various compounds. The functional characterization of active (net) transport requires that a larger number of experiments are performed than in the case of passive transport. Many variants of the experimental design can, therefore, be found in the literature. Some of the more straightforward experiments that have been used for the initial characterization of active net transport in epithelial cell monolayers are presented (points (a)–(c) are the same as those described for passive transport).

(a) Determination of transport rate in the apical to basal and basal to apical directions. The transport rate should be dependent on the transport direction in active transport.

(b) Determination of transport rate over a wide concentration interval (e.g. 10^{-6}–10^{-3} M). The transport of molecules should be saturable, i.e. the transport rate should be dependent on concentration.

(c) Determination of transport rate in the presence and absence of a general transport inhibitor such as low temperature (4 °C). The transport rate should decrease significantly at low temperatures if transport is active.

(d) Determination of transport rate against a concentration gradient.

(e) Inhibition of Na/K-ATPase by replacing sodium with choline and by the addition of ouabain. The sodium dependence of active transport in Caco-2 monolayers is often only partial.

(f) Inhibition of cellular metabolism. Only relatively small amounts of inhibitors such as sodium azide or 2-deoxyglucose can be used since ATP depletion also rapidly affects the integrity of the tight junctions. Inhibition of cellular metabolism is, therefore, relatively ineffective in Caco-2 monolayers.

(g) Inhibition with structural analogues or, when a well-characterized transport process is studied, with typical substrates for the carrier. Addition of the inhibitor from the apical and basolateral sides, respectively, may give information on the location of the carrier if the inhibitor is hydrophilic.

(h) pH-dependence.

Determination of active transport in Caco-2 monolayers also requires an understanding of the expression of the carrier with time in culture. Many carriers increase their expression in culture over time, and it is, therefore, important to use the monolayers at a time when the expression of the carrier is high and reproducible (39). This is particularly important with carriers displaying a lower expression in Caco-2 cells than *in vivo* (36). The type of carrier may also be altered with cellular differentiation (40). Finally, some carriers are proton-coupled and require a slightly acid pH for activity (41, 42). Therefore, investigators generally study active transport in Caco-2 monolayers under well-defined conditions. For instance, all studies are generally performed on cells of the same age.

Protocol 6 describes a method for the identification and initial characterization of active net transport across Caco-2 monolayers. For reasons of simplification, the method assumes that the compound of interest is available in radioactive form and that the active net transport is in the apical to basolateral direction. However, active transport can also be studied with unlabelled compounds using standard HPLC methods as discussed above. In this case, it may be necessary to concentrate larger samples than those described below in order to increase the sensitivity of the method.

Protocol 6. Characterization of active transport

Equipment and reagents

- Caco-2 cells cultivated in Transwell inserts as described in *Protocol 1*
- Radiochemical of interest
- Hank's Balanced Salt Solution (HBSS) (Gibco–BRL) supplemented with 25 mM Hepes, pH 7.4 (Sigma)

- HBSS/Mes: HBSS supplemented with 25 mM Mes, pH 6.0 (Sigma)
- Ouabain, sodium azide, 2-deoxyglucose (Sigma)
- Microtemp plate heater (Detrona)
- Liquid scintillation counter

A. *Identification of active transport*

1. Remove the complete cell culture medium and pre-incubate the cell monolayers with prewarmed HBSS for 30 min (see *Protocol 1*).
2. Prepare a 10^{-6} M solution of the radiolabelled compound of interest in prewarmed HBSS.
3. Add 0.4 ml of HBSS containing the compound of interest to the apical side and 1.5 ml of HBSS to the basolateral side of 3–4 monolayers. Take 10–50 μl samples from the apical chambers at $t = 0$ for determination of the initial radioactivity of each filter.
4. Transfer the inserts to new basolateral chambers containing fresh HBSS at 30 min intervals for 150 min. The choice of correct time interval is critical and may vary from one substance to another.
5. At the end of the experiment, take 10–50 μl samples from the apical chambers and 0.5–1.0 ml samples from the basolateral chambers.
6. Repeat steps 3 and 5, but add the drug to the basolateral chamber instead. Withdraw the samples from the apical receiving chamber and replace the medium in the apical chamber with fresh HBSS at regular time intervals.
7. Determine the radioactivity of the samples in a liquid scintillation counter.
8. Plot the cumulative amount of drug appearing in the basolateral chamber over time. Use linear regression analysis to calculate the flux rate J_{tot} (mol.min^{-1}.cm^{-2}) from the slope of the initial linear portion of the plot.
9. If a clear difference between the flux rates in the two opposite directions is observed, repeat steps 1–8, but use a 10^{-3} M concentration of the compound to investigate if the difference in flux rate is concentration-dependent. If no clear difference is observed, the transport may be passive or alternatively the expression and/or function of the carrier may not be optimal under the standard experimental conditions. In this case, it may be wise to study whether the carrier is dependent on, for example, the presence of protons, before the time-dependent expression of the carrier is investigated.

B. *Optimization of experimental conditions*

1. Investigate whether transport is dependent on the presence of protons by repeating the above studies with an apical pH of 6.0 using HBSS/Mes.

Protocol 6. *Continued*

2. Investigate the time-dependent polarity of the transport by repeating the above experiments in 5-, 10-, 15-, 20-, and 25-day-old monolayers. Perform the subsequent experiments with cells of an age displaying the best expression of vectorial transport.

C. *Further characterization of the active transport*

1. Investigate concentration-dependent transport by repeating steps 1–8 in the apical to basolateral direction at 37 °C at a concentration interval that allows the calculation of kinetic parameters.
2. Calculate the kinetic parameters with the use of a non-linear regression analysis program (e.g. PCNONLIN) according to the following expression:

$$J_{tot} = \frac{J_{max}\,C}{K_m + C} + P_d C;$$

where J_{tot} represents the total flux across the Caco-2 cell monolayer (mol.min^{-1}.cm^{-2} or mol.min^{-1}.(mg protein)$^{-1}$), J_{max} is the maximum transport rate for the carrier mediated process (mol.min^{-1}.cm^{-2}), C is the molar drug concentration on the donor side, K_m is the Michaelis–Menten molar constant, and P_d is the coefficient for the passive transport (cm min^{-1})

3. Investigate whether the transport occurs against a concentration gradient by studying the transport in the apical to basolateral direction in the presence of a 100-fold excess of unlabelled compound in the basolateral chamber.
4. Determine the sodium and energy dependence of the transport by repeating steps 1–7 with:
   (a) HBSS in which sodium chloride has been exchanged for choline chloride;
   (b) HBSS containing 100 μM of the Na/K-ATPase inhibitor ouabain; and
   (c) HBSS containing 1–5 mM sodium azide + 50 mM 2-deoxyglucose.
5. Determine the structural and stereo specificity by studying the transport in the apical to basolateral direction in the presence of a 50- to 100-fold excess of structural analogues, and by comparing the transport rates of D- and L-enantiomers of the compound.

3.3 Elimination of the unstirred water layer

For rapidly transported compounds, the monolayers should be stirred so that the effect of the 'unstirred water layer' is minimized (*Figure 3*). If a significant

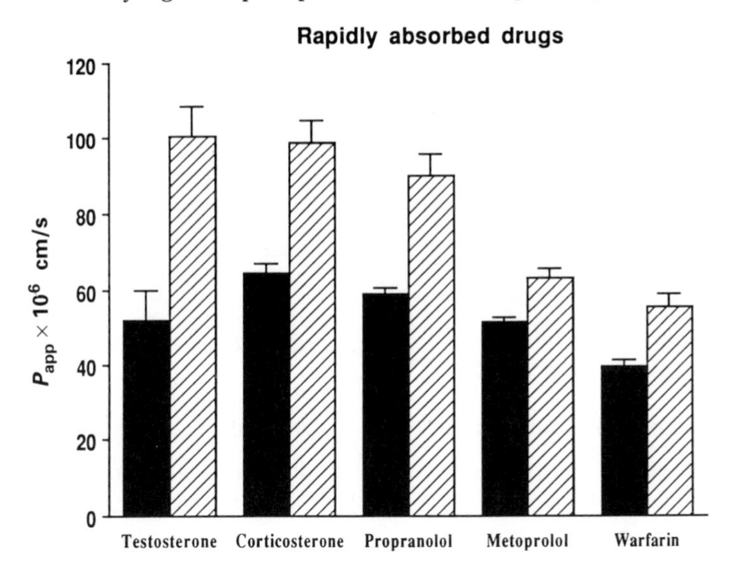

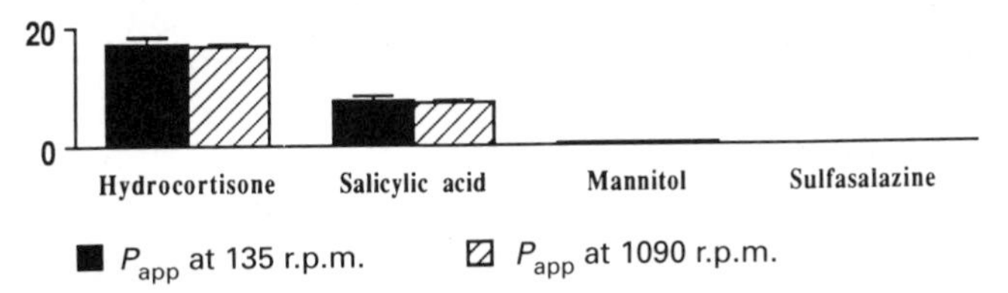

■ P_{app} at 135 r.p.m. ▨ P_{app} at 1090 r.p.m.

Figure 3. Influence of stirring on P_{app} for rapidly and slowly absorbed compounds in Caco-2 monolayers. Redrawn from ref. 44.

solute concentration difference between the liquid layer adjacent to the epithelial surface and the bulk occurs, the absorption rate will be erroneously low. Both passively and actively transported compounds may be affected by the 'unstirred water layer'. As can be seen in *Figure 3*, stirring is important for compounds having a $P_{app} > 2 \times 10^{-5}$/cm/s. Adequate stirring can be provided by a conventional ELISA-plate shaker or with a gas lift in a specially designed diffusion chamber (Costar). Note that vigorous stirring may affect the integrity of the monolayers. The quantitative method, presented in *Protocol 7*, for the determination of cell permeability coefficients (P_c) that are unbiased by the unstirred water layer should be used when a more precise determination of permeability is needed, for example when permeabilities *in vitro* are compared with those in the well-stirred *in vivo* situation (43, 44).

Protocol 7. Determination of permeability coefficients of the cell monolayer

Equipment and reagents

- Monolayers of Caco-2 cells cultivated in Transwell inserts (diameter 12 mm; pore size 0.4 μm) for 21–30 days (see *Protocol 1*)
- 12-well cell culture plates (Costar)
- HBSS containing the desired concentration of test compound
- HBSS (see *Protocol 2*)
- Plate shaker with stirring rate control (Titertec, Flow Laboratories)[a] placed in an incubator (37 °C and humidified air)
- Microtemp plate heater (Detrona)

Method

1. Equilibrate all solutions to 37 °C in a waterbath.
2. Remove the complete cell culture medium and pre-incubate four cell monolayers (in a cell culture plate) with HBSS both apically and basolaterally for 10–30 min at 37 °C.
3. Carefully pour or suction off the pre-incubation medium and place the cell monolayers in new cell culture wells containing 1.5 ml of HBSS. To start the experiment, add 0.4 ml of HBSS containing the desired concentration of test compound to the apical side of the cell monolayers. Take 10–50 μl samples from the apical chambers at $t = 0$ for determination of the initial concentration of each filter. Begin the stirring at a low speed (135 r.p.m.).
4. Stop the stirring at regular time intervals and transfer the Transwell inserts to new basolateral wells containing fresh HBSS. Routinely, three to six such transfers are made with each insert. It is important to choose sampling intervals such that sink conditions are maintained throughout the experiment, i.e. the concentration on the receiver side never exceeds 10% of the concentration on the donor side, especially when the test compound is rapidly transported.
5. At the end of the experiment, collect 10–50 μl samples from the apical side of the monolayers and 0.5–1.0 ml samples from the basolateral wells.
6. Repeat the experiment with four new cell monolayers at a higher stirring rate (767 r.p.m.). Alternatively, use the same cell monolayers and perform the experiment consecutively at two stirring rates (135 and 767 r.p.m.).
7. Determine the concentration of the samples and calculate the P_{app} (see *Protocol 2*) corresponding to each stirring rate.

[a] The stirring rates on this plate shaker were determined using a digital tachometer.

3.3.1 Calculation of P_c

The permeability coefficient of the cell monolayer is calculated using the following linear relationship between P_{app} and the corresponding stirring rates (43, 44):

$$\frac{V}{P_{app}} = \frac{1}{K} + \left(\frac{1}{P_c} + \frac{1}{P_f}\right)V; \qquad [1]$$

where: P_c = permeability coefficient of the cell monolayer; P_f = permeability coefficient of the filter support; V = stirring parameter (e.g. revolutions per minute, r.p.m.); K = is a lumped empirical constant.

This expression permits the determination of $(1P_c + 1/P_f)$ from the slope of the linear plot of V/P_{app} versus V.

As an example, P_{app} values for testosterone at 135 and 767 r.p.m. are $5.2 \pm 0.8 \times 10^{-5}$ and $9.6 \pm 0.3 \times 10^{-5}$ cm/s, respectively. With linear regression analysis of these data, Equation 1 becomes:

$$\frac{V}{P_{app}} = 1510 \times 10^3 + 8425V;$$

$$\text{slope} = \frac{1}{P_c} + \frac{1}{P_f} = 8425.$$

The permeability coefficient for the microporous polycarbonate filter can be calculated by:

$$P_f = \frac{D_{aq}\varepsilon}{h_f} \qquad [2]$$

where: D_{aq} is the aqueous diffusion coefficient of the drug (cm^2/s), and ε denotes the porosity and h_f the thickness of the filter.

In this example the dimensions of the filter are: $\varepsilon = 0.13$, $h_f = 10$ μm, and the aqueous diffusion coefficient of testosterone is 7.8×10^{-6} cm^2/s at 37 °C. The permeability coefficient thus becomes:

$$P_f = \frac{(7.8 \quad 10^{-6})\ (0.13)}{10 \times 10^{-4}} = 1014 \times 10^{-6} \text{ cm/s and } \frac{1}{P_f} = 986 \text{ s/cm}$$

It follows that the permeability coefficient of the cell monolayer becomes:

$$P_c = \frac{1}{8425 - 986} = 13.4 \times 10^{-5} \text{ cm/s};$$

and the relative standard deviation of P_c is equal to the relative standard deviation of the slope of Equation 1.

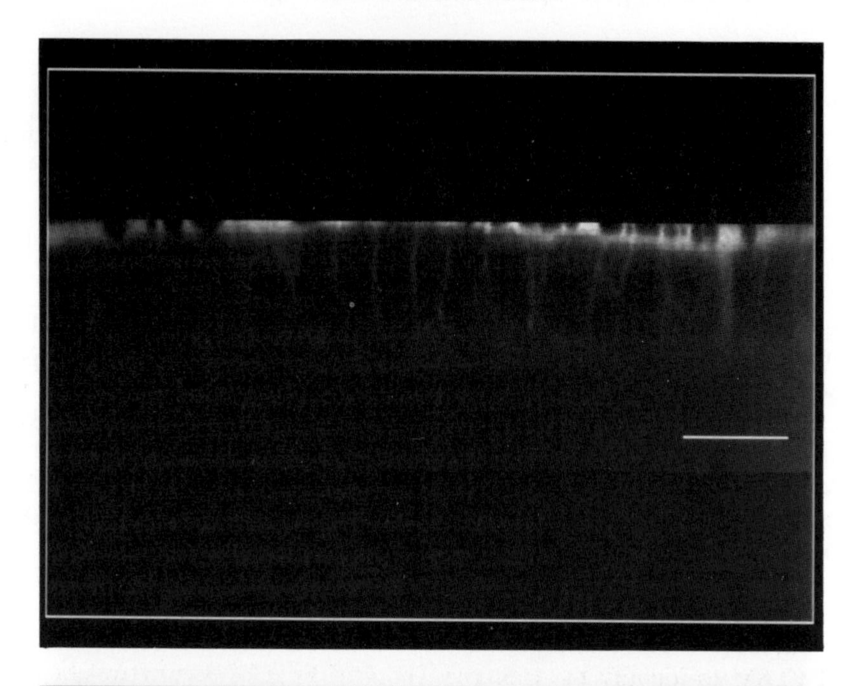

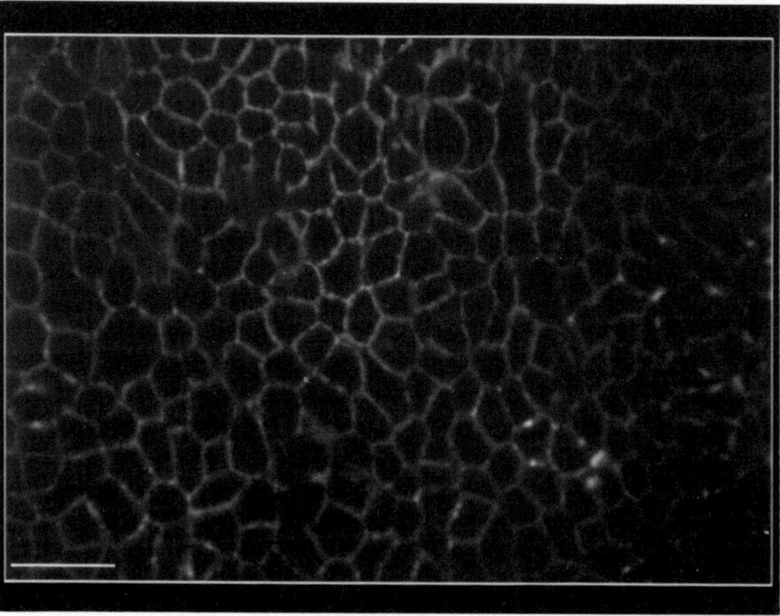

Figure 4. Confocal micrograph of a Caco-2 cell monolayer after a 60 min incubation with 1 mg/ml FD4. A, Optical horizontal cross-section. B, Optical vertical cross-section with step size in the z-direction of 0.24 μm. Top is apical, bottom is basolateral. The images have not been enhanced after acquisition.

When P_{app} is not influenced by stirring, transport is essentially controlled by the cell monolayer/filter. In this situation P_c is estimated by:

$$\frac{1}{P_c} = \frac{1}{P_{app}} - \frac{1}{P_f} \qquad [3]$$

3.4 Visualization of transcellular and paracellular permeability

As discussed previously, the transport of solutes across epithelia may occur via penetration of the intracellular space (passive diffusion, carrier-mediated transport, transcytosis) or via the intercellular space (1). Various methods can be used to determine the distribution of a compound in the epithelium and the transport pathways it has traversed. These include light and electron microscopic autoradiography, light and electron microscopic detection of electron-dense particles, and confocal laser scanning microscopy (CLSM) (45). CLSM has previously been used, for example, to localize transport pathways of fluorescence-labelled drugs in corneal tissue (46), oral mucosa (47), cultured cerebrovascular endothelium (48), and epithelial Caco-2 cell monolayers (49). The technique has been reviewed by Shotton (50). We have used CLSM to identify the transport route of large molecular weight and hydrophilic fluorescein isothiocyanate-labelled dextrans (FD) (M_w = 4000–20 000) in intact and manipulated Caco-2 cell monolayers. The method we used has been described by Hurni *et al.* (49), and was optimized for the detection of FDs. The use of other fluorescence-labelled model compounds requires a laser source and filter block applicable to the fluorescent probe under study. The optimum concentration of the fluorescence-labelled compounds depends on the permeability of the solute in the epithelial layer, the CLSM equipment, and the required sensitivity. Typical examples of CLSM pictures are given in *Figure 4A,B*. These show the distribution of fluorescence following a 60 min incubation of Caco-2 cell monolayers with FD4 (M_w = 4300). The hydrophilic model compound is observed in the extracellular space of the monolayer, while intracellular uptake of fluorescence cannot be detected. This indicates that FD traverses the epithelial cell monolayer mainly by the paracellular route.

4. Technical update

During the preparation of this article, a polyclonal rabbit anti-ZO-1 antibody (Zymed Laboratories, Inc) has appeared on the market which can be used on human fixed tissue. The antibody is raised against a 69 kD fusion protein (51) corresponding to amino acids 463–1109 of human ZO-1 cDNA lacking the alternatively spliced region (51, 52). The antibody reacts with human, mouse, rat, guinea pig, canine and chicken ZO-1. The antiserum can be used diluted

1:1000 on ethanol-fixed or formaldehyde-fixed, permeabilized Caco-2 cell monolayers.

References

1. Artursson, P. (1991). *Crit. Rev. Ther. Drug Carrier Syst.*, **8**, 305.
2. Chantret, I., Barbat, A., Dussaulx, E., Brattain, M. G., and Zweibaum, A. (1988). *Cancer Res.*, **48**, 1936.
3. Hidalgo, I. J., Raub, T. J., and Borchardt, R. T. (1989). *Gastroenterolgy*, **96**, 736.
4. Howell, S., Kenny, A. J., and Turner, A. J. (1992). *Biochem. J.*, **284**, 595.
5. Boulenc, X., Bourrie, M., Fabre, I., Roque, C., Joyeux, H., Berger, Y., and Fabre, G. (1992). *J. Pharmacol. Exp. Ther.*, **263**, 1471.
6. Peters, W. H. and Roelofs, H. M. (1992). *Cancer Res.*, **52**, 1886.
7. Baranczyk-Kuzma, A., Garren, J. A., Hidalgo, I. J., and Borchardt, R. T. (1991). *Life Sci.*, **49**, 1197.
8. Bjorge, S., Hamelehle, K. L., Homan, R., Rose, S. E., Turluck, D. A., and Wright, D. S. (1991). *Pharm. Res.*, **8**, 1441.
9. Hidalgo, I. J. and Borchardt, R. T. (1990). *Biochim. Biophys. Acta*, **1028**, 25.
10. Inui, K-I., Yamamoto, M., and Saito, H. (1992). *J. Pharmacol. Exp. Ther.*, **261**, 195.
11. Ng, K-Y. and Borchardt, R. T. (1993). *Life Sci.*, **53**, 1121.
12. Huntert, J., Jepson, M. A., Tsuruos, T., Simmons, N. L., and Hirst, B. H. (1993). *J. Biol. Chem.*, **268**, 14991.
13. Woodcock, S., Williamson, I., Hassan, I., and Mackay, M. (1991). *J. Cell. Sci.*, **98**, 323.
14. Dantzig, A. H., Hoskins, J., Tabas, L. B., Bright, S., Shepard, R. L., Jenkins, I. L., Duckworth, D. C., Sportsman, J. R., Mackensen, D., Rosteck Jr, P. R., and Skatrud, P. L. (1994). *Science*, **264**, 430.
15. Halleux, C. and Schneider, Y-J. (1991). *In Vitro Cell. Dev. Biol.*, **27A**, 293.
16. Hilgers, A. R., Conradi, R. A., and Burton, P. S. (1990). *Pharm. Res.*, **7**, 902.
17. Artursson, P. (1990). *J. Pharm. Sci.*, **79**, 476.
18. Candia, O., Mia, A. J., and Yorio, T. (1993). *Am. J. Physiol.*, **265**, C1479.
19. Tucker, S. P., Melsen, L. R., and Compans, R. W. (1992). *Eur. J. Cell Biol.*, **58**, 280.
20. Karnaky, J. K., Jr. (1992). In *Cell–cell interactions: a practical approach* (ed. B. R. Stevenson, W. J. Gallin, and D. L. Paul), pp. 257–74. IRL Press, Oxford.
21. Anderberg, E. K., Lindmark, T., and Artursson, P. (1993). *Pharm. Res.*, **10**, 857.
22. Artursson, P., Ungell, A-L., and Löfroth, J-E. (1993). *Pharm. Res.*, **10**, 1123.
23. Anderberg, E. K. and Artursson, P. (1994). In *Drug absorption enhancement: limitations, possibilities and trends* (ed. A. G. De Boer), pp. 101–18. Harwood Publishers, Chur, Switzerland.
24. Jones, K. H. and Senft, J. A. (1985). *J. Histochem. Cytochem.*, **33**, 77.
25. Anderberg, E. K. and Artursson, P. (1993). *J. Pharm. Sci.*, **82**, 392.
26. Madara, J. L. (1987). *Am. J. Physiol.*, **253**, C171.
27. Anderson, J. M., Stevenson, B. R., Jesaitis, L. A., Goodenough, D. A., and Mooseker, M. S. (1988). *J. Cell Biol.*, **106**, 1141.

28. Zahraoui, A., Joberty, G., Arpin, M., Fontaine, J. J., Hellio, R., Tavitian, A., and Louvard, D. A. (1993). *J. Cell Biol.*, **124**, 101.
29. Adams, C. J. and Storrie, B. (1981). *J. Histochem. Cytochem.*, **29**, 326.
30. McNeil, P. L. and Ito, S. (1989). *Gastroenterology*, **96**, 1238.
31. Somosy, Z., Kovacs, J., Siklos, L., and Köteles, G. J. (1993). *Scanning Microsc.*, **7**, 961.
32. Schultz, S. G. (1992). In *Essential medical physiology* (ed. L. R. Johnson), pp. 11–41. Raven Press, New York.
33. Karlsson, J., Kuo, S-M., Ziemniak, J., and Artursson, P. (1993). *Br. J. Pharmacol.*, **110**, 1009.
34. Ellis, J. A., Jackman, M. R., Perez, J. H., Mullock, B. M., and Luzio, J. P. (1992). In *Protein targeting: a practical approach* (ed. A. I. Magee and T. Wileman), pp. 187–215. IRL Press, Oxford.
35. Dharmsathaphorn, K. and Madara, J. L. (1990). In *Methods in enzymology* (ed. S. Fleischer, and B. Fleischer), Vol. 192, pp. 354–89. Academic Press, London.
36. Hu, M. and Borchardt, R. T. (1990). *Pharm. Res.*, **12**, 1313.
37. Cogburn, J. N., Donovan, M. G., and Schasteen, C. S. (1991). *Pharm. Res.*, **8**, 210.
38. Saito, H. and Inui, K-I. (1993). *Am. J. Physiol.*, **265**, G289.
39. Mahraoui, L., Rodolosse, A., Barbat, A., Dussaulx, E., Zweibaum, A., Rousset, M., and Brot-Laroche, E. (1994). *Biochem J.*, **298**, 629.
40. Wilson, G., Hassan, I. F., Dix, C. J., Williamson, I., Shah, R., Mackay, M., and Artursson, P. (1990). *J. Control. Release*, **11**, 25.
41. Dantzig, A. H. and Bergin, L. (1988). *Biochem. Biophys. Res. Commun.*, **155**, 1082.
42. Thwaites, D. T., Hirst, B. H., and Simmons, N. L. (1993). *Biochem. Biophys. Res. Commun.*, **1**, 432.
43. Karlsson, J. and Artursson, P. (1991). *Int. J. Pharmaceutics*, **71**, 55.
44. Karlsson, J. and Artursson, P. (1992). *Biochim. Biophys. Acta*, **1111**, 204.
45. Hoogstraate, A. J. and Boddé, H. E. (1993). *Adv. Drug Delivery Rev.*, **12**, 99.
46. Rojanasakul, Y., Paddock, S. W., and Robinson, J. R. (1990). *Int. J. Pharm.*, **61**, 163.
47. Hoogstraate, A. J., Cullander, C., Nagelkerke, J. F., Senel, S., Verhoef, J., Junginger, H. E., and Boddé, H. E. (1994). *Pharm. Res.*, **11**, 83.
48. Jaehde, U., Masereeuw, R., De Boer, A. G., Fricker, G., Nagelkerke, J. F., Vonderscher, J., and Breimer, D. D. (1994). *Pharm. Res.*, **11**, 442.
49. Hurni, M. A., Noach, A. B. J., Blom-Roosemalen, M. C. M., Nagelkerke, J. F., De Boer, A. G., and Breimer, D. D. (1993). *J. Pharmacol. Exp. Ther.*, **267**, 942.
50. Shotton, D. M. (1989). *J. Cell Sci.*, **94**, 175.
51. Willott, E., Balda, M. S., Heintzelman, M., Jameson, B., and Anderson, J. M. (1992). *Am. J. Physiol.*, **262**, C1119.
52. Willott, E., Balda, M. S., Fanning, A. S., Jameson, B., van Itallie, C., and Anderson, J. M. (1993). *Proc. Natl Acad. Sci. USA*, **90**, 7834.

7

A model of glandular epithelium for studying secretion

GÉRARD L. ADESSI, LAURENT BECK, and ABDERRAHIM MAHFOUDI

1. Introduction

The endometrium is a complex tissue consisting of the endometrial epithelium supported by a mesenchymal tissue called the endometrial stroma. The endometrial epithelium comprises the luminal epithelium, which faces the uterine lumen, and the glandular epithelium which invaginates into the endometrial stroma. The endometrium is a classic target tissue for ovarian steroids. It becomes a proliferative tissue under the action of oestrogens and, after ovulation, is changed by progesterone into a secretory tissue which accommodates ovum implantation. The cyclic morphological and functional differentiation of the whole endometrium under the effect of hormones represents the response of individual cell types within a complex signalling network.

The expression of tissue-specific functions in individual cells is governed at a fundamental level by the interactions of the cells with their environment (1, 2). Moreover, a prevailing concept of developmental biology holds that the interactions between an epithelium and the underlying mesenchyme mediate crucial aspects of normal development in embryonic (3, 4) as well as in adult tissues (5). These interactions are also believed to be important in neoplasia (6, 7). It has been reported that stroma mediates normal morphology and oestrogen responsiveness in cultured vaginal (8), uterine (8, 9), and mammary (10, 11) epithelia.

Although the glandular epithelium comprises only a minor fraction of the total endometrial tissue, glandular epithelial cells play an important role in mammalian reproduction by secreting fundamental biological products into the uterine lumen. By virtue of junctional complexes between the lateral plasma membranes of adjacent cells, endometrial glandular epithelial cells are immersed into two distinct microenvironments. The apical microenvironment contains a variety of epithelial secretory products, while the basal surface rests on a basement membrane through which basally secreted epithelial cell products may permeate and mix with blood-borne constituents and paracrine factors released by stromal cells.

The isolation and culture of endometrial cells is an indispensable experimental approach to the study of numerous aspects of endometrial cell biology. Although many investigators only maintain endometrial cells as short-term primary cultures (12–16), other's have serially cultured normal cells, particularly of epithelial origin (17, 18). In general, the epithelial cell monolayers obtained on solid plastic surfaces lose their *in vivo* hormone sensitivity. Furthermore the non-porous substrate precludes investigations of specialized, polarized functions such as transepithelial transport and vectorial secretion. A suitable *in vitro* model for studies of the differentiated state of glandular epithelial cells, epithelial–stromal interactions, and vectorial secretion, should:

(a) maintain the functional polarity of the endometrial cells;
(b) allow the separate analysis of the apical and basolateral secretions of epithelial cells; and
(c) allow epithelial and stromal cells to communicate without physically contacting each other.

Studies with mammary (19, 20), thyroid (21), pancreatic (22), and tracheal (23) epithelial cell cultures have shown that the maintenance of the differentiated state, which includes hormone sensitivity, requires the use of extracellular matrix extracts as substrata for the cells.

The natural basement membrane extracted from Engelbreth–Holm–Swarm mouse tumors (Matrigel; ref. 24), has been demonstrated to promote the differentiation of endometrial epithelial cells in primary cultures (25–28).

Using these studies as a basis, we have developed a method for subculturing highly homogeneous guinea-pig glandular epithelial cells on a Matrigel-coated permeable support in a serum-free, chemically defined medium (29, 30). These cells maintain their functional polarity and their responsiveness to hormones (29). The use of a serum-free medium in this model is an important approach as serum contains many substances which never come into contact with most body cells *in vivo* (31, 32) and can, therefore, modify the phenotypic behaviour of cells *in vitro* (33, 34). The use of a Matrigel-coated permeable support allows the determination of the vectorially-secreted (apical or basal) proteins (35).

This chapter describes first, the procedure for the isolation and culture of guinea-pig endometrial cells (epithelial and stromal), and second, the bicameral (two chamber) culture system in which epithelial cells are cultured on Matrigel-coated filters and stromal cells on solid plastic (see *Figure 1*).

2. Preparation

2.1 Cell culture media

During the development of the bicameral model to culture glandular epithelial cells, several media have been tested. Some of them are satisfactory both for primary cultures and subcultures: CMRL 1066, RPMI 1640 and Ham's F-12.

A *Endometrial cell isolation*

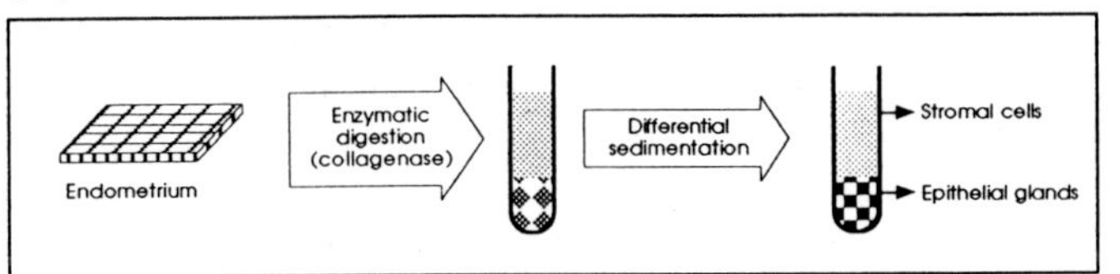

B *Culture steps and bicameral culture system*

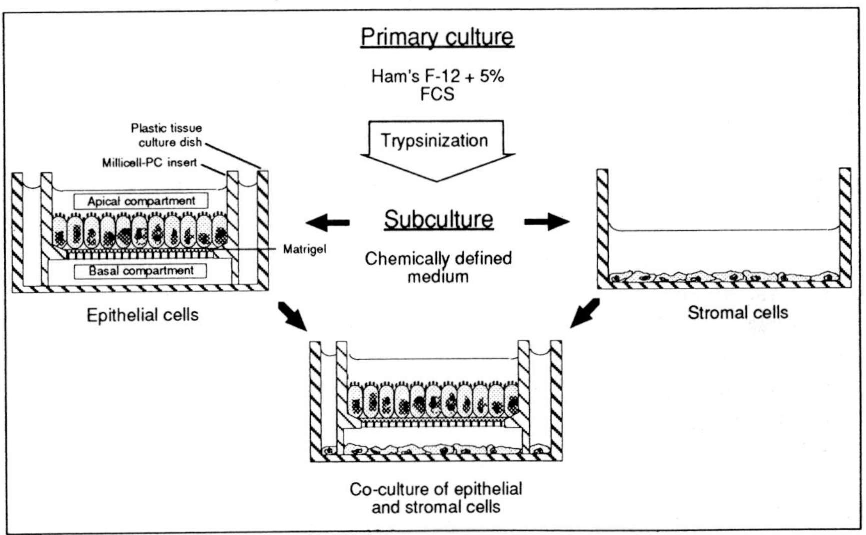

Figure 1. Endometrial cell culture. (A) Representation of the different steps taken to isolate the endometrial epithelial glands from their surrounding stroma; (B) Culture steps and schematic representation of the bicameral culture system. After a primary culture step, glandular epithelial cells are seeded on Matrigel-coated filters and grown to confluence, preventing any mixture between basal and apical compartments. In this bicameral system, stromal cells, subcultured separately, can be included in the basal compartment to study the modulation of epithelial secretion by the underlying homologous stroma.

Phenol red-free media should be used to prevent the oestrogenic effect of this compound. Ham's F-12 supplemented with 5% FCS induces a good cell proliferation when cells are cultured in tissue culture flasks. Ham's F-12 without FCS is suitable for the subculture of glandular epithelial cells on Matrigel-coated filters. This medium is used as the basic component of the chemically defined medium (see *Table 1*)

2.2 Animals

Uteri should be taken from virgin mature female guinea-pigs of the Hartley albino variety (500–600 g bodyweight), fed *ad libitum* with standard food

Table 1. Endometrial cell culture media

Media	**Preparation of media**
Hanks' Balanced Salt Solution, Ca^{2+} and Mg^{2+} free (CMF-HBSS)	The powder used is a modified Hanks' Balanced Salt Solution (Sigma, cat. no. H-4891): without calcium chloride, magnesium sulfate, phenol red, and sodium bicarbonate. This solution is prepared following the supplier's instructions, buffered at pH 7.3 with 0.35 g/litre sodium bicarbonate and 20 mM Hepes, sterilized immediately by filtration (0.22 μm porosity) and stored at 4 °C. Prior to use, this solution is supplemented with: penicillin 100 IU/ml streptomycin 100 μg/ml fungizone 2.5 μg/ml
Basal medium (BM)	Sodium bicarbonate and phenol red-free Ham's F-12 (Gibco–BRL). This medium is supplemented with 1.176 g/litre of sodium bicarbonate and 20 mM Hepes when prepared (pH 7.3).
Serum-supplemented medium (SSM)	BM with bovine insulin (10 μg/ml), Na_2SO_4 (600 μM), penicillin (100 IU/ml), streptomycin (100 μg/ml), fungizone (2.5 μg/ml), and 5% (v/v) FCS.
Serum-free, chemically defined medium (CDM)	SSM, devoid of FCS, and supplemented with EGF (10 ng/ml), transferrin (10 μg/ml), sodium selenite (10 ng/ml), retinol acetate (100 ng/ml), and BSA (1 mg/ml).
Labelling medium (LM)	Modified CDM prepared with methionine-free Ham's F-12 (Gibco–BRL) supplemented with 0.6 μg/ml methionine, and containing a reduced amount of BSA (100 μg/ml).

enriched with ascorbic acid. The animals should be housed under controlled temperature (20°C) and lighting conditions (12 h light:12 h dark). The oestrous cycle is controlled by vaginal opening (duration of the oestrous cycle: 15–16 days). Only those females which display at least two successive regular oestrous cycles should be used.

A detailed procedure for the isolation of guinea-pig endometrium is described in *Protocol 1*. Three or four pooled uteri are routinely used yielding approximately 5×10^6 glandular epithelial cells uterus.

Protocol 1. Isolation of guinea-pig endometrium[a]

Equipment and reagents[b]

- CMF-HBSS: prepare as described in *Table 1* with the addition of 50 μg/ml gentamycin
- 10 cm diameter sterile plastic tissue culture dishes (Falcon, cat. no. 3003)
- 10 ml sterile plastic pipettes (Falcon)
- 40 ml sterile plastic can (Falcon)
- Sterile plastic scraper (Falcon, cat. no. 3085)
- Binocular magnifying glass (Olympus)
- Sterile scissors, forceps, and scalpels (Moria)
- 70% (v/v) ethanol in tap water
- Female guinea-pigs (see Section 2.2)
- Laminar air-flow hood

Method

1. Kill the animals by decapitation on the day of vaginal opening.[c]
2. Immobilize the animal on its back and soak the entire abdomen with 70% ethanol.
3. Make a midline incision, push the intestinal contents aside and expose the genital tractus.
4. Excise the uterus. Remove any extraneous tissue if present and place it in a 40 ml sterile can.
5. Conduct the remaining procedures under sterile conditions within a horizontal laminar air-flow hood. Carry out all manipulations in 10 cm diameter culture dishes containing 6–8 ml of cold, sterile CMF-HBSS.
6. Transfer the uterus to a dish and separate the two horns using a scalpel.
7. Using forceps, tear off the myometrium under binocular glass examination (7 ×).[d] Transfer the endometrium to a new dish.
8. Make an incision lengthways with scissors to open the endometrium and cut it crosswise into two equal parts.
9. Scrape off the surface of the uterus lumen using a plastic scraper to remove any mucus. Transfer the endometrial parts little by little into a new dish.
10. Gently striate a culture dish by making a grid pattern with a scalpel, and add 2–3 ml of CMF-HBSS.
11. Place the endometrium on the striated dish and press it on to the bottom of the dish. The grid pattern prevents the tissue from slipping.
12. Mince the endometrium with a scalpel into 2.5 mm^3 cubes, transfer them to a 40 ml plastic can, and wash the fragments extensively with sterile CMF-HBSS until the supernatant is clear. Aspirate and discard the liquid.

[a] It takes an experienced operator 1½–2 h to process 3 animals.
[b] Glassware and stable solutions are autoclaved at 121°C and labile solutions are filtered through a 0.22 μm porosity membrane filter (Millipore; Millex GV, cat. no. SLGV 025 BS). Tap

Protocol 1. *Continued*

water is first prefiltered, and then subjected to reverse osmosis, mixed-bed deionization, and submicron filtration. Plastics and commercially available media are already sterilized and used without treatment.

[c] We routinely use a home-made guillotine rather than large scissors to kill animals. Heavily etherized animals can be used.

[d] The myometrium is torn from the endometrium like a sock by turning out one end and stripping it along the longitudinal axis of the horn. The endometrium appears as a pipe in the central part of the horn.

2.3 Matrigel-coated filters

Epithelial cells generally adhere, spread, and differentiate more satisfactorily on biomatrix substrata than on inert tissue culture filters. We have investigated different procedures for coating filters including using poly-D-lysine, fibronectin, FCS, and Matrigel (29). Our best results, in terms of cell adhesion and differentiation, have been obtained using the latter (24). *Protocol 2* details the method of coating Millicell filters with Matrigel.

Protocol 2. Coating filters with Matrigel

Equipment and reagents

- Millicell-PC culture filter inserts (pore size 3 µm, diameter 30 mm) (Millipore, cat. no. PITT 030 50)
- Pipettes, plates, and tubes (Falcon). Precool at −20 °C.
- Gentamicin solution (10 mg/ml; Sigma, cat. no. G-1272)
- Matrigel: phenol red-free (Falcon, cat. no. 40234–0005). Thaw the Matrigel overnight at 4°C. Dilute the Matrigel 1:5 in ice-cold Ham's F-12 media (BM, see *Table 1*) containing 50 µg/ml gentamicin[a]
- 6-well tissue culture plates (Falcon, cat. no. 3046)
- Laminar vertical air-flow hood

Method

1. Conduct the protocol on ice and within the sterile atmosphere of a vertical (down-flow) laminar air-flow hood.
2. Place the filters in a 6-well plate. Overlay each filter with 250 µl of the diluted Matrigel.
3. Rock the filters gently to spread the solution uniformly over the entire surface of the filters.
4. Warm the coated filters for 1 h at 37 °C.
5. Aspirate any excess liquid and air-dry the Matrigel overnight under sterile conditions at room temperature.
6. Irradiate the dried gel for 30 min using UV light.[b]

[a] Diluted Matrigel can be stored for one month at 4 °C.

[b] Matrigel-coated filters can be stored for one week at room temperature under sterile conditions.

3. Isolation and culture of glandular epithelial cells

3.1 Isolation of glandular epithelial cells

Large and dense epithelial glands are present in the guinea-pig endometrium during the pre-oestrous and the oestrous phases. To take advantage of this fact, females are killed on the day of vaginal opening. Glands in the endometrium fragments are easily dissociated from the surrounding tissue by careful treatment with collagenase. During this treatment, the glands must remain intact. This can be monitored using inverted microscopy. The glands are isolated from dissociated cells (mainly stromal cells and fibroblasts) and undigested fragments by gentle differential gravitation sedimentation.

Various collagenase preparations are commercially available for tissue culture studies and each differs in its collagenase activity and the presence of other contaminating enzymes including proteases and clostripain. Consequently, each batch of collagenase needs to be tested before use and the procedure to achieve a good glandular dissociation adjusted accordingly. In order to optimize the conditions for digestion, it is preferable to reduce the collagenase concentration and alter the incubation time whilst monitoring the digestion process. *Protocol 3*, which must be conducted in the sterile atmosphere of a vertical, laminar air-flow hood, details the procedure for isolating and culturing glandular epithelial cells from guinea-pig endometrium.

Protocol 3. Isolation and primary culture of glandular epithelial cells

Equipment and reagents

- Vertical laminar air-flow hood
- Inverted phase-contrast microscope (e.g. Olympus TV 1)
- Incubator (Flow Laboratories)
- 6 and 10 cm diameter tissue culture dishes (Falcon, cat. no. 3004 and 3003)
- 25 and 75 cm^2 tissue culture flasks (Falcon, cat. no. 3013E and 3024)
- 50 ml centrifuge tubes (Falcon, cat. no. 2070)
- 14 ml plastic tubes (Falcon, cat. no. 2057)
- Collagenase solution: 0.25% (w/v) collagenase (CLS I Worthington from Seromed) dissolved prior to use in CMF–HBSS (see *Table 1*) and sterilized by filtration[a]
- FCS (Gibco–BRL)
- HF: CMF–HBSS supplemented with 1% (v/v) FCS
- SSM (see *Table 1*)
- Endometrial fragments from three guinea-pig uteri (see *Protocol 1*)

A. *Isolation of epithelial glands*

1. Conduct the whole procedure in a vertical, laminar air-flow hood. Transfer the endometrial fragments to a 10 cm culture dish.
2. Add 30 ml of collagenase solution and incubate at 37 °C for 1 h.
3. Examine the digestate with an inverted phase-contrast microscope after 1 h of incubation and then every 15 min. Handle the dish with care to prevent the fragments sticking to each other. Usually the

Protocol 3. *Continued*

majority of the glands are dissociated from the surrounding tissue after 120 min. However, depending on the activity of the collagenase and animal variability, the incubation time may vary from 90–150 min. The digestate contains isolated glands, free cells, and undigested endometrial fragments.

4. Transfer the digestate into a 50 ml centrifuge tube, make the volume up to 45 ml with HF, and turn the tube upside down to resuspend the digestate (see *Figure 2*).
5. Pellet the endometrium fragments by allowing them to settle under normal gravity for 30 s–1 min, and collect two-thirds of the supernatant by aspiration.[b]
6. Transfer this supernatant in 10 ml fractions into three 14 ml plastic tubes, and allow them to pellet under normal gravity.
7. Make up the volume of the centrifuge tube (obtained at step 5) to 45 ml with HF[c] and repeat steps 5–7.
8. Stop the procedure (steps 4–7) when the pellet in the 50 ml centrifuge tube primarily contains undigested fragments of endometrium.
9. Aspirate the supernatants from the 14 ml plastic tubes and pool the pellets two by two. Retain the supernatants for the culture of stromal cells (see *Protocol 7*).
10. Resuspend the pellet in 10 ml of HF and pellet under normal gravity.
11. Repeat steps 9 and 10 as necessary to finally obtain one pellet in one 14 ml tube (see *Figure 2*).
12. Wash the final pellet (containing undisrupted epithelial glands) with 10 ml of HF. Resuspend the glands and pellet under normal gravity.
13. Repeat step 12 three or more times in order to obtain a suspension of epithelial glands devoid of cellular fragments of stromal origin.
14. Resuspend the final pellet in 12 ml of SSM.

B. *Primary culture*

1. Seed 1 ml of the epithelial gland suspension into one 25 cm^2 tissue culture flask and add 3 ml of SSM.
2. Assess the purity and density of the seeded glands and adjust the volume of the original gland suspension accordingly to obtain a density of approximately 20 glands/cm^2.[d]
3. Incubate the epithelial glands at 37 °C in a humidified atmosphere of 5% CO_2 in air.[e]

4. Change the SSM 24 h after seeding and renew the culture medium at 48 h intervals until the glandular epithelial cells achieve 70–80% confluence. This should take 6–8 days.[f]

[a] The collagenase solution should be filtered through a 0.8/0.2 μm porosity double membrane filter (Gelman Acrodisk PF, cat. no. 4187).
[b] The supernatant contains epithelial glands of various sizes, fragments of stroma, vessels, and agglomerated and isolated cells.
[c] Take care to add the liquid rapidly and vigorously in order to resuspend the pellet.
[d] Starting with three uteri, enough material should be obtained to seed twelve 25 cm^2 tissue culture flasks.
[e] During primary culture, the epithelial glands rapidly attach to the plastic surface within the first 2 h of incubation. After 2 h they spread out and the peripheral cells actively proliferate. Eventually the cells from neighbouring glands meet to form a monolayer.
[f] In our laboratory, functional primary cultured glandular epithelial cells survive for 2–3 weeks with the medium renewed every 2 days.

3.2 Morphology of endometrial cells *in vitro*

The aspect and growth characteristics of endometrial cells in primary culture are shown in *Figure 3*. After cell isolation, the epithelial component is made of glandular organoids (*Protocol 3B*, step 2) composed of several hundred epithelial cells (10 μm in diameter; see *Figure 3a*), whereas the stromal component (*Protocol 7A*, step 2) is composed of isolated cells (5–15 μm in diameter) which attach to the plastic surface within 30 min (see *Figure 3b*).

During the first 48 h following plating, the epithelial glands progressively adhere to the surface of the flask (see *Figure 3c*). The plating efficiency depends on the integrity of the glands after digestion by collagenase. The morphology of the stromal cells is mostly fibroblastic and they rapidly reach confluence (see *Figure 3d*). After 5 days, epithelial cells grow as monolayers composed of cohesive groups of polygonal cells (*Protocol 3B*, step 4; see *Figure 3e*).

3.3 Subculture of glandular epithelial cells into the bicameral system

Glandular epithelial cells are harvested for subculture when they reach 70–80% confluence in 25 cm^2 tissue culture flasks. Primary cultures are subjected to selective trypsinization in the presence of EDTA (see *Protocol 4A*). A two-step procedure is adopted: the first step encourages the rapid dissociation of contaminating non-epithelial cells, i.e. stromal cells and fibroblasts; the second step allows the dissociation of glandular epithelial cells. In order to reduce enzymatic damage to the cell surface, the contact between the cells and the enzyme solution has to be minimized. Once the epithelial cells are dissociated, they are reseeded on to Matrigel-coated filters in the bicameral system (see *Protocol 4B*).

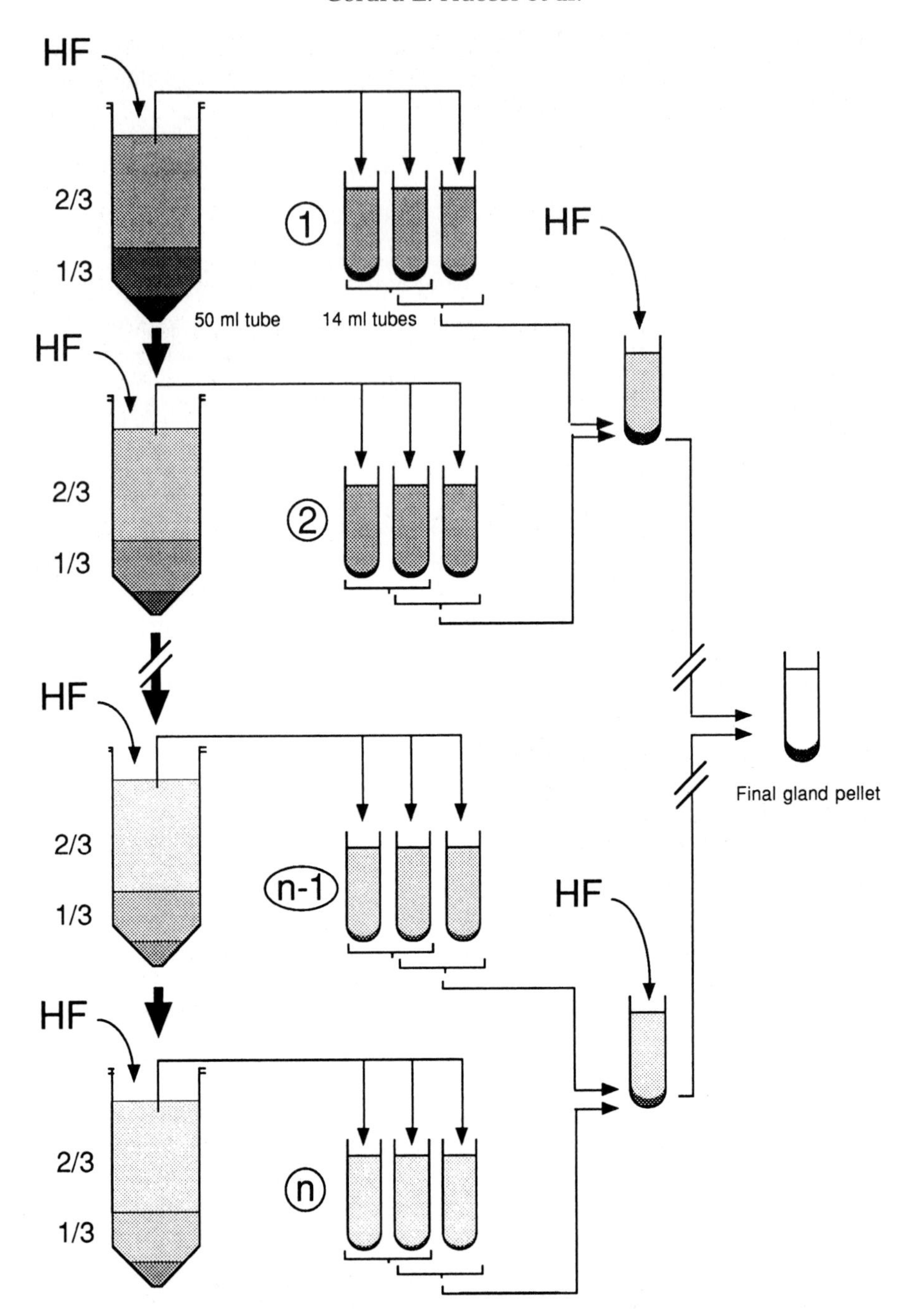

Figure 2. Schematic representation of how to collect epithelial glands using differential sedimentation by gravitation. (*n*, number of times the supernatant is transferred from the 50 ml tube to the 14 ml tube; usually *n* is about 8–10.)

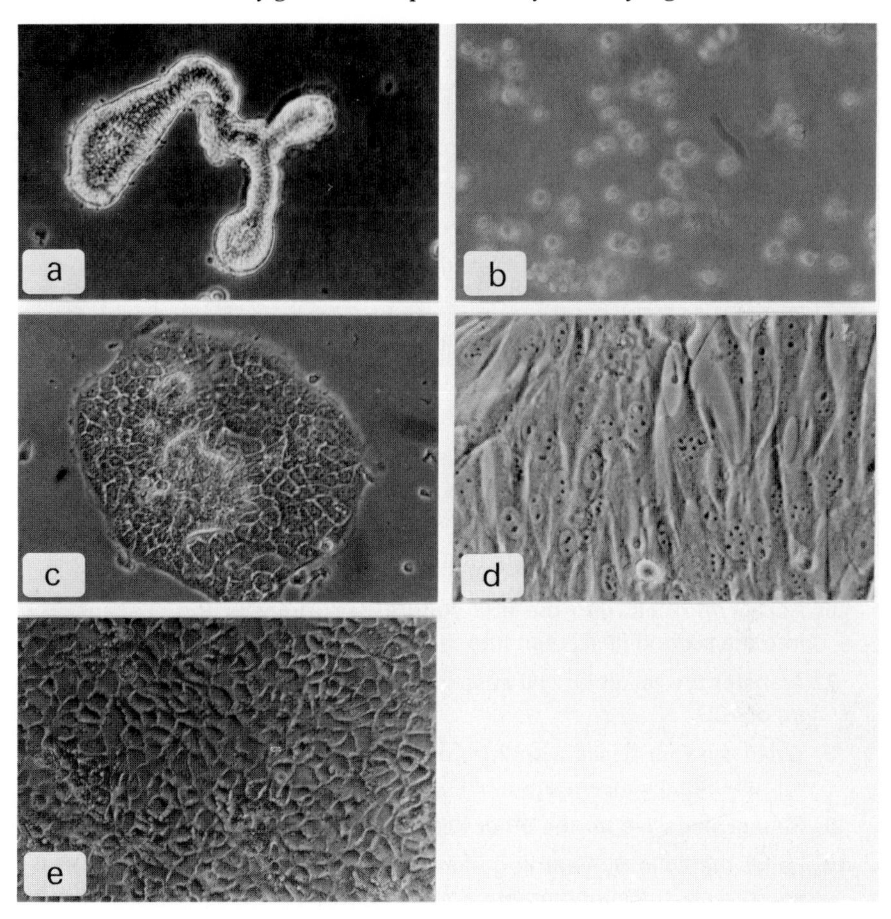

Figure 3. Phase-contrast photographs of the two endometrial cell types in culture. (a) An epithelial gland immediately after cell isolation; (b) stromal cells immediately after cell isolation; (c) an epithelial gland adhering to the bottom of the flask after 48 h in culture medium; (d) stromal cells at 48 h; (e) epithelial cells on days of culture growing as monolayers composed of cohesive groups of polygonal cells.

Protocol 4. Subculture of glandular epithelial cells on to Matrigel-coated filters in the bicameral system

Equipment and reagents

- Vertical, laminar air-flow hood
- Inverted phase-contrast microscope (Olympus TV 1)
- Incubator (Flow Laboratories)
- 14 ml plastic tubes (Falcon, cat. no. 2057)
- 10 ml sterile pipettes (Falcon)
- Subconfluent cultures of endometrial glandular epithelial cells in 25 cm^2 flask (see *Protocol 3B*, step 4)
- 6-well tissue culture plates (Falcon, cat. no. 3046) and Matrigel-coated Millicell-PC filter inserts (see *Protocol 2*)

Protocol 4. *Continued*

- Haemocytometer
- HE: CMF-HBSS (see *Table 1*) supplemented with 0.02% (w/v) EDTA
- HET: HE supplemented with 0.1% (w/v) trypsin (Sigma, cat. no. T-0646)
- CDM (see *Table 1*)
- 0.08% Trypan Blue: prepare 0.08% (w/v) Trypan Blue (Flow) in saline and filter it through a 0.22 μm pore size filter
- FCS

A. *Cell dissociation*

1. Conduct the whole protocol in a vertical, laminar air-flow hood. Remove and discard the culture media from four 25 cm^2 flasks containing the epithelial cells. Treat the flasks four at a time.
2. Wash the cells twice for 1 min, each time, at 25°C with 2 ml of HE. Discard the washings.
3. Incubate the cells for 1 min at room temperature with 2 ml of HET/flask.
4. Aspirate off the HET.
5. Incubate the cells for 10 min at 37°C with 1 ml of HET/flask.
6. Add 3 ml of HE into the first culture flask, transfer the suspension into the second flask, then into the third and finally the fourth.
7. Transfer the resulting cell suspension into a 14 ml tube containing 1 ml of FCS.[a]
8. Wash the four flasks with 2 ml of HE and pool the liquid into the 14 ml tube (final concentration of FCS = 10% (v/v)).
9. Repeat steps 1–8 for the other flasks.
10. Pellet the cells by centrifugation (75 *g*, 10 min, room temperature) and discard the supernatants.
11. Resuspend the pellets and pool them in 10 ml of CDM. Pellet the cells by centrifugation (75 *g*, 10 min, room temperature).

B. *Seeding and subculture*

1. Discard the supernatant and resuspend the final pellet of glandular epithelial cells into 10 ml of CDM.
2. Add 2 ml of CDM to each well of the tissue culture plates and place a Matrigel-coated filter insert into each well to form the bicameral system (see *Figure 1B*).
3. Determine the percentage cell viability.[b]
4. Dilute the cell suspension to 5.6×10^5 viable cells/ml with CDM and seed 1.5 ml aliquots on to the surface of each Matrigel-coated filter: cell density = 2×10^5 viable cells/cm^2.
5. Incubate the bicameral system at 37°C in a humidified atmosphere of 5% CO_2 in air.

6. Change the medium every 48 h as follows:
   (a) add 2 ml of CDM to each well of a new plate;
   (b) discard the media from the apical compartments of each insert;
   (c) place the Matrigel-coated filters into the new wells;
   (d) add 1.5 ml of CDM to the apical compartments.
7. Incubate the bicameral system for 4–5 days in order to obtain a confluent monolayer of epithelial cells.[c]

[a] FCS is required to inhibit the trypsin activity.
[b] Dilute 0.1 ml of cell suspension 1:10 with culture medium. Mix 0.1 ml of the diluted solution with 0.1 ml of 0.08% Trypan Blue and count the stained (non-viable) and unstained (viable) cells with a haemocytometer.
[c] Confluence can be monitored by either inverted phase-contrast microscopy or by measuring the transepithelial electrical resistance (TER; see Section 4).

4. Study of cellular and vectorially secreted proteins by glandular epithelial cells

In order to identify protein markers of cellular and vectorially secreted proteins, glandular epithelial cells cultured in the bicameral system are labelled with [^{35}S]methionine. (see *Protocol 5*). The radiolabelled proteins extracted from the cells (cellular proteins) or released into the apical or basal compartments (secreted proteins) are analysed by sodium dodecyl sulfate polyacrylamide gel electrophoresis (SDS-PAGE) and fluorography.

Only confluent cultures of glandular epithelial cells on Matrigel-coated filters are used for metabolic labelling. Confluence is monitored routinely by examination under an inverted phase-contrast microscope. The confluence of filter-cultured glandular epithelial cells is confirmed by measuring the TER before use. The resistance to the flow of ions across the monolayer is an indication that transepithelial permeability barriers have become established (25, 36, 37; see Chapter 3, Sections 1 and 3). Transepithelial resistance can be measured using a Millicel-ERS voltohmmeter (Millipore). The electrical resistance of the filter-cultured glandular epithelial cells progressively increases through the first three days and then plateaus for the remaining period of culture (until the 10th day).

Protocol 5. Study of cellular and vectorially-secreted proteins

Equipment and reagents

- Inverted phase-contrast microscope (e.g. Olympus TV 1)
- Incubator (Flow Laboratories)
- Centrifuge (Beckman)
- 15 ml glass centrifuge tubes (Corex)
- 10 ml sterile pipettes (Falcon)
- 6-well plastic tissue culture plates (Falcon, cat. no. 3046)

Protocol 5. *Continued*

- L-methionine in PBS: dissolve 4.48 mg/litre L-methionine in PBS. Keep on ice
- LM (see *Table 1*)
- [^{35}S]methionine: Sp. Act. 1000 Ci/mmol (Amersham). Dilute to 100 μCi/1.7 ml in LM
- Lysis solution: 9 M urea, 3% (v/v) Nonidet P-40, 0.5% (w/v) sodium deoxycholate, 2% (v/v) β-mercaptoethanol, and 2 mM phenylmethylsulfonyl fluoride (PMSF) in distilled water
- Bicameral system containing a confluent monolayer of glandular epithelial cells on Matrigel-coated filters (see *Protocol 4*)
- Laemmli buffer 2 ×: 0.125 M Tris–HCl, 4% (w/v), SDS, 10% (v/v) glycerol, and 10% (v/v) β-mercaptoethanol in distilled water (pH 6.8)
- Acetone and 80% (v/v) acetone in distilled water

A. *Metabolic labelling*

1. Add 1.7 ml of LM containing 100 μCi [^{35}S]methionine to each well of a new plate.
2. Discard the apical media from each filter insert.
3. Rinse the filter from both apical and basal sides with LM without [^{35}S]methionine.
4. Place the filters in each well of a new plate.
5. Add 0.5 ml of LM without [^{35}S]methionine into the apical compartment of the bicameral system.
6. Incubate the bicameral system at 37 °C in a humidified atmosphere of 5% CO_2 in air for 6 h.
7. Collect the media in each apical compartment at the end of the labelling period. Rinse the apical compartments with 0.5 ml of LM devoid of radioactive tracer, and pool the liquids.
8. Take out the inserts and collect the labelled media in the basal compartments.

B. *Preparation of [^{35}S]methionine-labelled cellular proteins*

1. After the labelling period and the collection of the incubation media, briefly wash the cells in cold PBS containing L-methionine.
2. Add cold PBS containing L-methionine to the cells and maintain them at 4 °C for 10 min.
3. Discard the L-methionine solution and wash the cells twice more with PBS containing L-methionine.
4. Add 150 μl of lysis solution to each apical compartment and incubate them at room temperature for 15 min.
5. Collect the lysates add an equal volume of 2 × Laemmli buffer, and store at – 80 °C until further processing.

C. *Preparation of [^{35}S]methionine-labelled secreted proteins*[a]

1. Centrifuge the media at 4 °C for 15 min at 6000 *g*.

2. Collect the supernatant, add 5 volumes of acetone, and keep overnight at −20 °C.
3. Centrifuge at −20 °C for 15 min at 6000 *g*.
4. Wash the pellet twice with 80% acetone at −20 °C.
5. Dry the pellet under nitrogen.
6. Add 50 µl of lysis solution and an equal volume of 2 × Laemmli buffer to the dry pellet. Store at −80 °C until further processing.

[a] The same procedure is used for the labelled media collected from either the apical or basal compartments (see *Protocol 5A*).

Cellular and secreted proteins are analysed by SDS-PAGE and fluorography following standard methods (see Chapters 2 and 3). A typical result for proteins secreted preferentially to the apical or basal compartments by the filter-cultured glandular epithelial cells is given in *Figure 4.* A conservative analysis identified the protein bands which appeared to be secreted in a polarized manner as the high molecular mass proteins (apical) and the 110, 32, 21.5, 18.5, 15.5, and 13 kDa proteins (basal). Some proteins secreted in both compartments are more intensely labelled in the apical compartment, such as the 44 kDa protein band, or in the basal compartment, such as the 180 and 14 kDa protein bands.

5. Study of the effect of progesterone on proteins secreted into both the apical and basal compartments

The bicameral system allows the study of the effect of progesterone on proteins secreted into both the apical and basal compartments by glandular epithelial cells.

5.1 Hormonal treatment

When epithelial glandular cells are at confluence (*Protocol 4B*, step 7) they are incubated for 48 h in CDM with or without steroid hormones. Steroid hormones dissolved in ethanol (10^{-2} M) are added to the incubation medium with a final concentration of 0.01% (v/v) ethanol (2 ml of medium in the basal compartment and 1.5 ml in the apical compartment). The cells cultured in the bicameral system receive hormones from both basal and apical reservoirs.

Oestrogen-stimulated cells are incubated for 48 h in CDM containing 10^{-8} M oestradiol (E_2). Progesterone-stimulated cells are primed with 10^{-8} M.E_2 for 48 h and 5×10^{-8} M progesterone is added to the priming medium for the

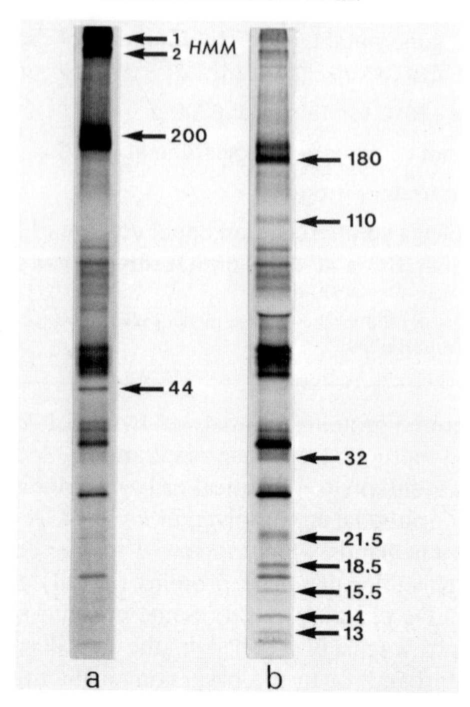

Figure 4. Electrophoretic profiles of [^{35}S]methionine-labelled secreted proteins on SDS-PAGE. The secreted proteins were obtained at confluence from either the apical (lane a) or the basal (lane b) compartments of the bicameral system, glandular epithelial cells cultured on Matrigel-coated filters in chemically defined medium. Their apparent molecular masses were determined from the molecular mass standards run concurrently. HMM, high molecular mass proteins.

last 16 h. The untreated cells (control) receive vehicle alone: 0.01% (v/v) ethanol. At the end of the incubation period both apical and basal compartments are washed with CDM or methionine-free CDM, depending on the next treatments.

5.2 Progesterone receptor content of glandular epithelial cells

Confluent glandular epithelial cells grown on Matrigel-coated filters contain progesterone receptors. We routinely use progesterone receptor immunostaining to assess the progesterone responsiveness of these cells. The protocol has been described by Perrot-Applanat *et al.* (38), and this method is used with some modifications as reported by Chaminadas *et al.* (39; see *Protocol 6*).

Protocol 6. Progesterone receptor immunostaining

Equipment and reagents

- Zeiss reticular micrometer (10 × 10 mm square)
- 10 ml pipettes (Falcon)
- Mouse monoclonal antibody mPR II[a] (6–25 μg IgG/ml PBS)
- Peroxidase-conjugated anti-mouse IgG antibody (DAKO); use a 1:50 dilution in PBS
- DAB stain: 0.5% 3, 3′-diaminobenzidine (DAB; Sigma) and 0.02% hydrogen peroxide in 50 mM Tris–HCl buffer (pH 7.6)
- PBS, pH 7.3
- FCS (Gibco)
- Picric acid–paraformaldehyde: add 20 g of paraformaldehyde to 150 ml of saturated picric acid in distilled water. Heat to 65 °C in a fume hood. Add several drops of 1 M NaOH to clear the solution, then filter and allow it to cool. Make up to 1 litre with PBS
- Methanol (ice-cold)

Method

1. Fix the cells in picric acid–paraformaldehyde, postfix in methanol, and treat with 5% (v/v) FCS in PBS to minimize the non-specific binding of reagents in subsequent steps.
2. Remove the excess FCS by washing three times in PBS, then incubate the cells for 18 h at 4 °C with a non-specific antibody or with mouse monoclonal antibody mPR II.
3. Wash the cells three times with PBS and then incubate with the peroxidase-conjugated anti-mouse IgG antibody for 1 h at room temperature in a humidified chamber.
4. Wash the cells three times with PBS.
5. Incubate the cells with DAB stain for 1–2 min at room temperature.
6. Determine the number of labelled nuclei with the micrometer at 250 × magnification, in a set of four randomly chosen confluent subcultures.
7. Express the results as the percentage of nuclei which are immunostained.

[a] Mouse monoclonal antibody raised against guinea-pig progesterone receptors (mPR II) is purchased from Transbio.

Typical results are displayed in *Figure 5*. After the first 4 days of culture, the immunostained nuclei averaged 50%. The immunostaining was restricted to the nuclei and no significant variation in the percentage of immunostained nuclei was observed in cultured cells in the presence of 10^{-8} M E_2 prolonged for 48 h. When 10^{-7} M progesterone was added after 4 days of culture in CDM containing 10^{-8} ME_2, a significant decrease in immunostained nuclei was observed after 24 h and no immunostained nuclei were detected after 36 h. Consequently, further studies should be conducted within the 24 h following progesterone addition when the percentage of cells exhibiting progesterone receptors remains high.

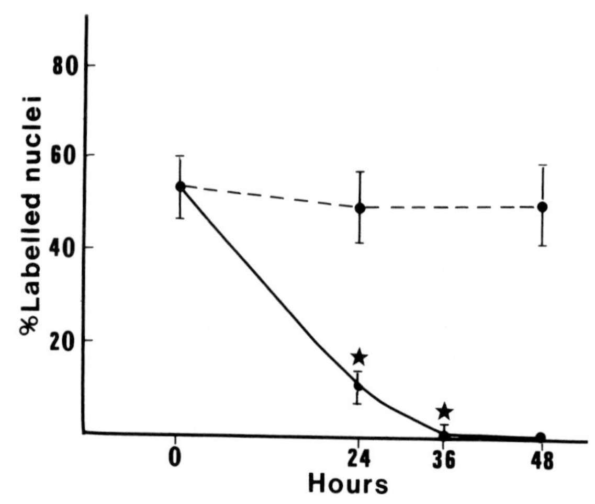

Figure 5. Progesterone effect on the percentage of progesterone receptors in glandular epithelial cells. The percentage of immunostained nuclei was calculated in confluent (t_0) and post-confluent (24, 36, and 48 h) cultures on Matrigel-coated dishes in CDM. 10^{-7} M progesterone (solid line) or vehicle (dotted line) were added at t_0 to the culture medium containing oestradiol (E_2; 10^{-8} M). Data are the mean ± SD from quadruplicate cultures. ★ $p < 0.05$ versus E_2 treatment.

5.3 Effect of progesterone on vectorially-secreted proteins

To study the effect of progesterone on vectorially secreted proteins, glandular epithelial cells are treated with E_2 or E_2 plus progesterone, as described in Section 5.1. The cells are then labelled with [^{35}S]methionine according to *Protocol 5*. A representative result is displayed in *Figure 6*. Compared with E_2-treated glandular epithelial cells, progesterone acting in combination with E_2 induces, in both the apical and basal compartments, a decrease in the labelling of the 88 and 53 kDa protein bands and an increase in the labelling of the 28 kDa protein band. The labelling intensity of the 28 kDa protein band increases much more in the basal than in the apical compartment under progesterone treatment. In the apical compartment, progesterone induces the secretion of a 137 kDa protein band.

6. Co-culture of glandular epithelial cells with stromal cells in the bicameral system

The interactions between the epithelium and the underlying mesenchyme mediate crucial aspects of normal tissue development. The endometrium is an epithelium/mesenchyme stromal structure responsive to circulating steroid hormones. However, this tissue has an active role in producing autocrine and paracrine factors, e.g. growth factors and cytokines. These factors are pro-

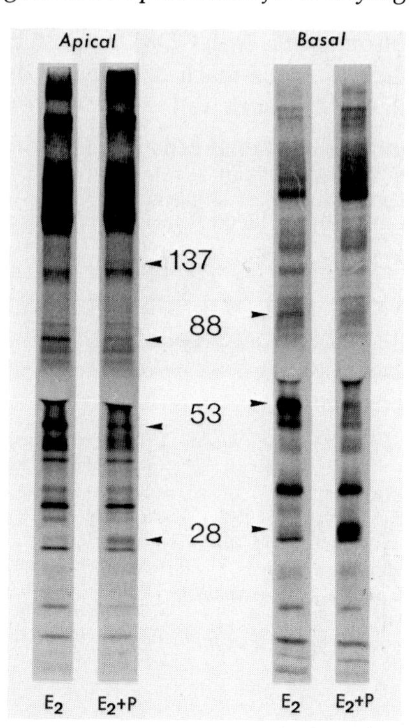

Figure 6. Progesterone effect on the electrophoretic profiles of secreted proteins. Glandular epithelial cells subcultured on Matrigel-coated filters were treated with 10^{-8} M E_2 or E_2 plus 5×10^{-8} M progesterone (P) for 16 h. Compounds altered by progesterone are indicated by their apparent molecular weight. (Apical, proteins secreted into the apical compartment; basal, proteins secreted into the basal compartment.)

duced by either the epithelial or stromal cells, and control the proliferation and differentiation of the different cell types of the endometrium (9, 40, 41). Particularly in the endometrium, stromal cells mediate normal morphology and oestrogen responsiveness of epithelial cells (8, 9). A rigorous evaluation of such tissue–tissue interactions in the endometrium requires an improved *in vitro* system to that already described. This model should maintain the functional polarity of the epithelial cells while permitting the separate analysis of the apical and basolateral secretions of epithelial cells, and the secretions of the stromal cells. Organ cultures or mixed stromal–epithelial cultures are useful models (9, 42) but are limited in their use because although morphological changes can be observed, the secretory products of the endometrial cell population cannot be distinguished.

A bicameral culture system allows the study of the interactions between glandular epithelial cells cultured on Matrigel-coated filters, and stromal cells grown on the culture surface of the basal compartment. To constitute this co-culture system (see *Figure 1B*) the following procedure is adopted:

(a) Subculture the stromal cells with CDM in 6-well tissue culture plates (see *Protocol 7*). When the cells reach confluence, discard the medium and add 2 ml of fresh CDM to each well.

(b) Culture the glandular epithelial cells in CDM on Matrigel-coated filters (see *Protocol 4*). When the cells reach confluence, discard the medium.

(c) Place the filters into the wells containing the stromal cells.

(d) Add 1.5 ml of CDM to the apical compartment.

Protocol 7. Culture of stromal cells

Equipment and reagents

- Inverted phase-contrast microscope (Olympus TV 1)
- Incubator (Flow Laboratories)
- 14 ml plastic tubes (Falcon, cat. no. 2057)
- 50 ml centrifuge tubes (Falcon, cat. no. 2070)
- 10 and 50 ml sterile pipettes (Falcon)
- 75 cm^2 plastic tissue culture flasks (Falcon, cat. no. 3024)
- 6-well plastic tissue culture plates (Falcon, cat. no. 3046)
- HF: CMF–HBSS (see *Table 1*) supplemented with 1% FCS (Gibco)
- SSM (see *Table 1*)
- HE: CMF–HBSS supplemented with 0.02% (w/v) EDTA
- HET: HE supplemented with 0.1% (w/v) trypsin (Sigma, cat. no. T-0646)
- CDM (see *Table 1*)
- Trypan Blue (Flow Laboratories)
- FCS

A. *Primary culture of stromal cells*

1. Centrifuge the supernatant obtained in *Protocol 3A,* step 9 at 25 *g* for 10 min.[a]
2. Recover the supernatant, discard the pellet (containing cell aggregates) and centrifuge the supernatant at 100 *g* for 10 min.
3. Resuspend this second pellet with 18 ml of SSM.
4. Seed the stromal cells using 2 ml of suspension in each of the 75 cm^2 tissue culture flasks, and add 10 ml of SSM/flask (about 2×10^5 cells/cm^2).
5. Incubate the flasks at 37 °C in a humidified atmosphere of 5% CO_2 in air.
6. Renew the medium every 2 days. After 6–7 days the cells should be 70–80% confluent.

B. *Cell dissociation*

1. Remove and discard the culture medium from three 75 cm^2 flasks.
2. Wash the primary stromal cell cultures twice with 5 ml of HE, and discard the washings.
3. Incubate the cells for 10 min at 37 °C with 2 ml of HET.
4. Add 2 ml of HE into the first culture flask. Transfer the suspension into the second and then the third flask.

5. Transfer the resultant cell suspension into a 14 ml tube containing 1 ml of FCS.
6. Wash the three flasks with 2 ml of HE and pool the washings into the 14 ml tube.
7. Repeat steps 1–6 for the other flasks.
8. Pellet the cells by centrifugation (75 *g*, 10 min) and discard the supernatants.
9. Resuspend the pellets and pool them in 10 ml of CDM. Pellet the cells by centrifugation (75 *g*, 10 min).

C. *Subculture*

1. Discard the supernatant and resuspend the final pellet in 10 ml of SSM.
2. Determine the number of viable cells using the Trypan Blue exclusion test (*Protocol 4*, footnote *b* therein).
3. Make adequate cell dilutions in CDM to seed aliquots (2 ml) into 6-well tissue culture plates with a viable cell density of 10^5 cells/cm^2.[b]
4. Incubate the plates at 37 °C in a humidified atmosphere of 5% CO_2 in air.
5. Discard the tissue medium after 24 h and add 2.5 ml of CDM. Renew the medium every 48 h until confluence is achieved (within 4–6 days).

[a] The supernatants are pooled and treated as fractions of approximately 50 ml volume.
[b] Cell density may vary from 0.5–3 × 10^5 cells.

We use the bicameral system to study the effect of stromal cells on the proteins secreted into the apical compartment by progesterone-treated glandular epithelial cells. Stromal cells are cultured in the basal compartment and confluent epithelial cells are grown on the Matrigel-coated filter insert. Under these conditions the labelling of the 137 kDa protein band of the apical compartment increased 2.8-fold in progesterone-treated cells as compared with epithelial cells cultured without stromal cells (see *Figure 7*). Such an effect would suggest a paracrine influence of stromal cells through secreted short-lived signal molecules.

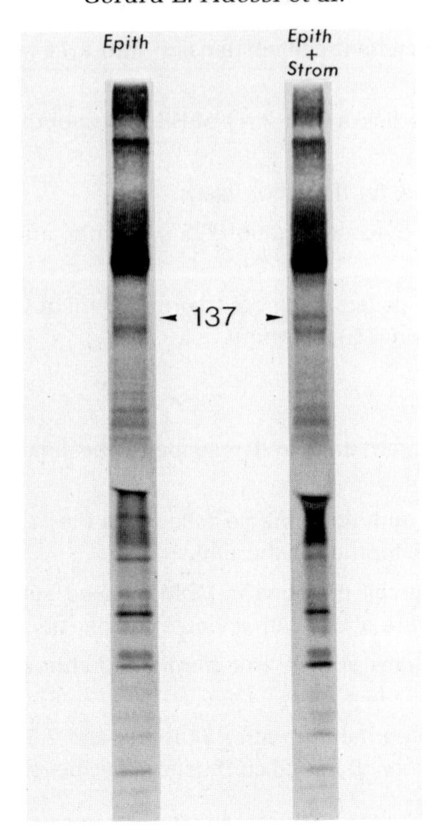

Figure 7. Modulation by the 'stroma' of electrophoretic profiles of apical secreted proteins. Glandular epithelial cells cultured on Matrigel-coated filters alone (Epith) or with stromal cells (Epith + Strom) were treated with 10^{-8} M E_2 plus 5×10^{-8} M progesterone for 16 h. Compounds altered by progesterone treatment are indicated in the centre.

References

1. Bernfield, M., Banerjee, S., Rapraeger, A., Jalkanen, M., Koda, J., Nguyen H., and Kaznowski, C. (1984). *J. Cell. Biol.*, **99**, 234.
2. Bissell, M. J. and Aggeler, J. (1987). In *Mechanisms of signal transduction by hormones and growth factors* (ed. 1), p. 251. Alan R. Liss, Inc., New York.
3. Cunha, G. R. (1976). *J. Exp. Zool.*, **196**, 361.
4. Kratochwil, K. (1986). *Am. J. Pathol.*, **149**, 23.
5. Cunha, G. R., Bigsby, R. M., Cooke, P. S., and Sugimura, Y. (1985). *Cell Diff.*, **17**, 137.
6. Camps, J. L., Chang, S. M., Husu, T. C., Freeman, M. R., Hong, S. J., Zhau, H. E., VonEschenbach, A. C., and Chung, L. W. K. (1990). *Proc. Natl Acad. Sci. USA*, **87**, 75.
7. Miller, F. R., McEacherm, D., and Miller, B. E. (1989). *Cancer Res.*, **49**, 6091.
8. Cooke, P. S., Uchima, F. D. A., Fujii, D. K., Bern, H. A., and Cunha, G. R. (1986). *Proc. Natl Acad. Sci. USA*, **83**, 2109.

9. Inaba, T., Wiest, W. G., Strickler, R. C., and Moni, J. (1988). *Endocrinology*, **123**, 2355.
10. Howlett, A. R. and Bissell, M. J. (1990). *Protoplasma*, **159**, 85.
11. McGrath, C. M. (1983). *Cancer Res.*, **43**, 1355.
12. Bongso, A., Gajra, B., Lian, N. P., Wong, P. C., Soon-Chye, N., and Ratman, S. (1988). *Hum. Reprod.*, **3**, 705.
13. Chaminadas, G., Propper, A. Y., Royez, M., Prost, O., Rémy-Martin, J. P., and Adessi, G. L. (1986). *J. Reprod. Fertil.*, **77**, 547.
14. Kirk, D., King, R. J. B., Heyes, J., Peachet, L., Hirsch, P. J., and Taylor, R. W. T. (1978). *In Vitro*, **14**, 651.
15. Satyaswaroop, P. G., Bressler, R. S., De La Pena, M. M., and Gurpide, E. (1979). *J. Clin. Endocrinol. Metab.*, **48**, 639.
16. Varma, V. A., Melin, S. A., Adamec, T. A., Dorman, B. H., Siegfried, J. M., Walton, L. A., Carmey, C. N., Morton, C. R., and Kaufman, D. G. (1982). *In Vitro*, **18**, 911.
17. Gentola, G. M., Cisar, M., and Knab, D. R. (1984). *In Vitro Cell Dev. Biol.*, **20**, 451.
18. Trent, J. M., Davis, J. R., and Payne, C. M. (1980). *Am. J. Obstet. Gynecol.*, **136**, 356.
19. Emmerman, J. T., Burnwen, S. J., and Pitelka, D. R. (1979). *Tissue Cell*, **241**, 87.
20. Parry, G., Cullen, B., Kaetzel, C. S., Kramer, R., and Moss, L. (1987). *J. Cell. Biol.*, **105**, 2043.
21. Chambard, M., Cambion, J., and Mauchamps, J. (1981). *J. Cell. Biol.*, **91**, 157.
22. Hootman, S. R. and Logsdon, C. D. (1988). *In Vitro Cell Dev. Biol.*, **24**, 566.
23. Benali, R., Dupuit, F., Jaquot, J., Fuchey, C., Hinnrasky, J., Ploton, D., and Puchelle, E. (1989). *Biol. Cell*, **66**, 263.
24. Kleinman, H. K., McGarvey, M. L., Hassel, J. R., Star, V. I., Cannon, F. B., Laurie, G. W., and Martin, G. R. (1986). *Biochemistry*, **25**, 312.
25. Glasser, S. R., Julian, J., Decker, G. L., Tang, J. P., and Carson, D. D. (1988). *J. Cell. Biol.*, **107**, 2409.
26. Negami, A. I. and Tominaga, T. (1989). *Hum. Reprod.*, **4**, 620.
27. Rinehart, C. A., Lyn-Cook, B. D., and Kaufman, D. G. (1988). *In Vitro Cell. Dev. Biol.*, **24**, 1037.
28. White, T. E. K., Di Sant'Agnese, P. A., and Miller, R. K. (1990). *In Vitro Cell. Dev. Biol.*, **26**, 636.
29. Mahfoudi A., Nicollier, M., Propper, A. Y., Coumes-Marquet, S., and Adessi, G. L. (1991). *Biol. Cell*, **71**, 255.
30. Mahfoudi, A., Fauconnet, S., Bride, J., Beck, L., Remy-Martin, J. P., Nicollier, M., and Adessi, G. L. (1992). *Biol. Cell*, **74**, 255.
31. Clark, J. (1983). In *Hormonally defined media: a tool for cell biology* (ed. G. Fischer and R. J. Wieser), p. 6. Springer–Verlag, Berlin Heidelberg.
32. Ham, R. G. (1983). In *Hormonally defined media: a tool for cell biology* (ed. G. Fischer and R. J. Weiser), p. 16. Springer–Verlag, Berlin, Heidelberg.
33. Medrano, E. E., Resnicoff, M., Cafferata, E. G. A., Larcher, F., Podhajcer, O., Bover, L., and Molinari, B. (1990). *Exp. Cell Res.*, **188**, 2.
34. Orly, J. (1984). In *Cell culture methods for molecular and cell biology.* (ed. D. W. Barnes, D. A. Sirbasku, and G. H. Sato), Vol. 2, p. 63. Alan R. Liss, New York.
35. Mahfoudi, A., Nicollier, M., Beck, L., Mularoni, A., Cypriani, B., Fauconnet, S., and Adessi, G. L. (1994). *J. Reprod. Fertil.*, **100**, 637.

36. Cereijido, M., Robbins, E. S., Dolan, W. J., Rotonno, C. A., and Sabatini, D. D. (1978). *J. Cell Biol.*, **77**, 853.
37. Hadley, M. A., Djakiev, D., Byiers, S. W., and Dym, M. (1987). *Endocrinology*, **120**, 1097.
38. Perrot-Applanat, M., Logeat, F., Groyer-Picard, M. T., and Milgrom, E. (1985). *Endocrinology*, **116**, 1473.
39. Chaminadas, G., Rémy-Martin, J. P., Alkhalaf, M., Propper, A. Y., and Adessi, G. L. (1989). *Cell Tissue Res.*, **257**, 129.
40. Brigstock, D. R., Heap, R. B., and Brown, K. D. (1989). *J. Reprod. Fertil.*, **85**, 747.
41. Tabibzadeh, S. (1991). *Endocrinol. Rev.*, **12**, 272.
42. Hohn, H. P., Winterhagen, E., Busch, L. C., Mareel, M. M., and Denker, H. W. (1989). *Cell Tissue Res.*, **257**, 505.

8

A model of the blood–testis barrier for studying testicular toxicity

ANNA STEINBERGER and ANDRZEJ JAKUBOWIAK

1. Introduction

Although diverse classes of chemicals have been found to produce spermatogenic damage in experimental animals, their primary target site(s) and/or mechanisms of action are, in most cases, not known (1). Certain compounds, e.g. diester phthalates, need to be metabolized to monoesters in the liver or gut before they can exert a toxic effect on the testis (2). A toxicant may affect the male gametes *indirectly* by interfering with the production of androgens by Leydig cells, or with various functions of Sertoli cells, e.g. metabolism, vectorial secretion, and the formation of specialized tight junctions. As Sertoli cells are frequently affected by exposure to testicular toxicants, they have been most widely utilized in studies of testicular toxicity *in vitro*.

During prepubertal testicular maturation, neighbouring Sertoli cells develop highly specialized tight junctions that divide the seminiferous epithelium into a basal and adluminal compartment. These junctional complexes provide the structural basis for the so-called 'blood–testis' barrier which restricts free passage of macromolecules into the adluminal compartment. Thus, germ cells located in the adluminal compartment (spermatocytes and spermatids) are differentiating in a unique microenvironment maintained by both the selective permeability of the 'blood–testis' barrier, and the Sertoli cell secretions. The integrity of the blood–testis barrier is believed to be critical for the normal spermatogenic process (3). Until recently, studies of the blood–testis barrier could only be conducted *in vivo*. However, the development of two-compartment Sertoli cell cultures (4–6) has provided a very useful model for exploring the formation of Sertoli cell tight junctions (7) and their responses to regulatory agents and environmental xenobiotics *in vitro* (8, 9). Under appropriate conditions, Sertoli cells in two-compartment cultures become morphologically and functionally polarized and in many respects resemble their counterparts *in vivo* (see *Figure 1*; refs 4–6, 10). However, it should be emphasized that results from *in vitro* experiments should not be extrapolated directly to whole animals, although they may provide

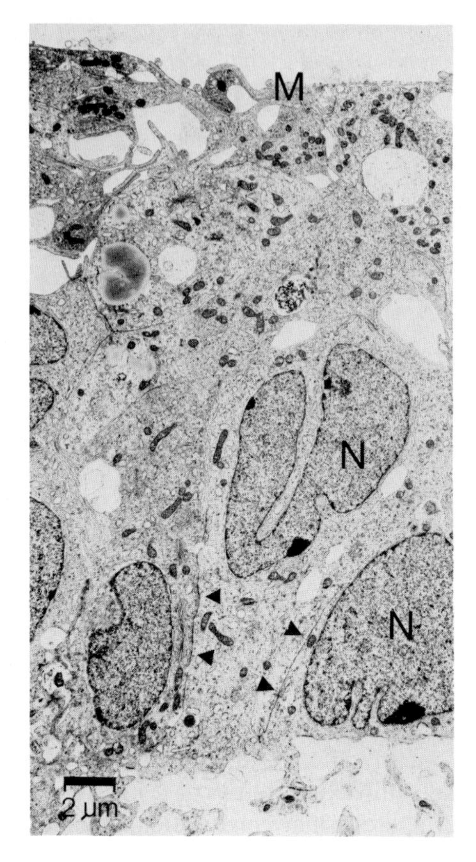

Figure 1. Ultrastructural features of immature rat Sertoli cells cultured on a permeable membrane in a two-compartment chamber system. Note the polarized morphology, characteristic nuclei (N) with infolded nuclear membrane and prominent nucleoli, and the presence of specialized tight junctions (arrowheads) between adjacent Sertoli cells. The apical cytoplasm forms numerous microvilli (M) (reproduced from A. Janecki and A. Steinberger, *Endocrinology*, Vol. 120, 1987, with permission of The Endocrine Society).

important information on the *direct* effects and mechanisms by which toxicants can cause testicular damage.

2. Two-compartment cultures of Sertoli cells

2.1 Advantages and potential pitfalls

Two-compartment cultures (see *Figure 2*) have many advantages over cultures of Sertoli cells on glass or plastic:

(a) The cells are maintained on permeable supports which can be coated with reconstituted extracellular matrix (e.g. Matrigel; see Chapter 7,

Protocol 2) in order to mimic the basement membrane of seminiferous tubules.

(b) In contrast to the squamous appearance of cells cultured on solid supports, Sertoli cells maintained on permeable membranes in the two-compartment culture system form a highly polarized columnar epithelial layer with intercellular tight junctions similar to those observed *in vivo* (see *Figure 1*).

(c) As *in vivo*, the Sertoli cells derive nutrients through the basolateral membrane, and secrete their products bidirectionally.

(d) The two-compartment culture system permits the investigation of the effects of testicular toxicants on the formation and maintenance of Sertoli cell tight junctions. Examples of toxicological studies using two-compartment Sertoli cell cultures are described in Section 3.

(e) The role of cellular interactions on the expression of testicular toxicity can be investigated using co-cultures of Sertoli cells with germ cells, peritubular myoid cells, and/or Leydig cells (see *Figure 2b*).

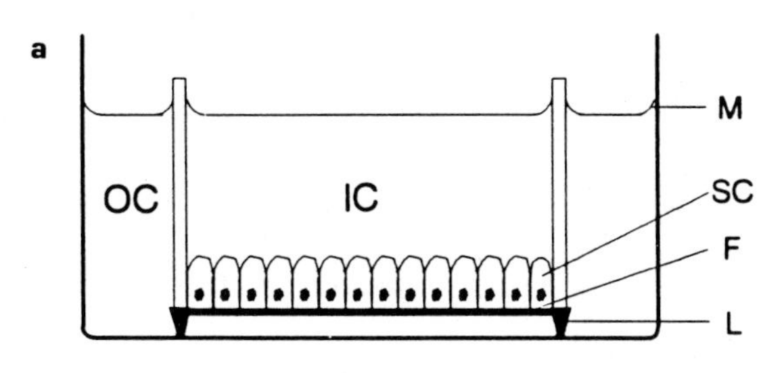

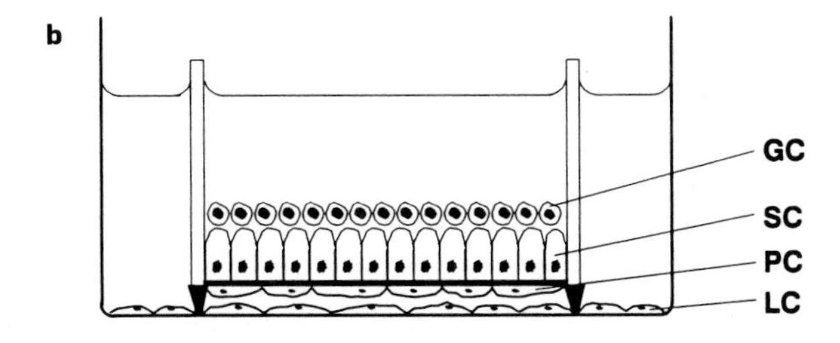

Figure 2. The two-compartment culture system. Sterile filter inserts (prepared as described in *Protocol 1*) placed into a 12-well plastic culture dish form inner (IC) and outer chambers (OC), respectively. (a) The Sertoli cells (SC) are seeded on top of the supporting filter (F), which can be coated with extracellular matrix prior to use. Appropriate medium (M) is added to the outer chambers. (b) Co-culture of Sertoli cells (SC) with peritubular myoid cells (PC), Leydig cells (LC), and germ cells (GC) for the study of paracrine interactions between various testicular cell types. The Leydig cells are usually plated on a layer of extracellular matrix (e.g. Matrigel) and, for the first 24 h, in medium containing 2% fetal bovine serum. The germ cells can be placed either directly on top of the Sertoli cells or on an additional filter to avoid direct physical contact with the Sertoli cells (reproduced from A. Steinberger and J.-P. Clinton, *Methods in toxicology*, Vol. 3A, 1993, with permission of Academic Press, Inc.).

Two-compartment cultures of Sertoli cells also have a number of potential pitfalls which must be considered in order to avoid misinterpretation of experimental results:

(a) Since Sertoli cells cease to divide around the time of puberty (15–18 days in the rat), the cell density at the time of plating is critical for the formation of a confluent monolayer essential for neighbouring cells to form tight junctions. The integrity of the cell monolayer and tight junctions can be assessed, for example, by determining either the permeability of the monolayer to insulin (see Section 2.4.1) or the transepithelial electrical resistance (TER; see Section 2.4.2). The presence of an incomplete monolayer would lead to the erroneous interpretation of the experimental results.

(b) The measurements of TER are subject to considerable variation due to various factors, including the pH, temperature, and the position of the electrodes, and therefore must be conducted under strictly standardized conditions.

(c) The age of the donor animals can profoundly influence the behaviour of Sertoli cells in culture. Thus, results obtained with cells from a particular age may not be applicable to other ages.

(d) If secretory proteins are to be measured, non-specific binding of the proteins to the filter (especially those made of nitrocellulose) must be minimized by coating the filters prior to use, e.g. with albumin.

2.2 Construction of the culture chambers

Although the two-chamber assembly is commercially available, we have used 12-well tissue culture plates for the outer chambers and prepared our own inner chambers (see *Protocol 1*). Construction of the culture chambers is relatively simple and cost-effective as the glass part of the chamber can be recycled many times.

Protocol 1. Construction of the two-chamber culture system

Equipment and reagents

- 12-well tissue culture plates (Linbro, cat. no. 76–053–05; ICN Flow)
- Pyrex tubing (18 mm i.d., 20 mm o.d.; Corning, cat. no. 230177). Cut the tubing into 9 mm lengths using a glass cutter[a]
- 7X detergent (Linbro, cat. no. 76–670–94; ICN Flow)
- 732 RTV-sealant (Bearing Inc., cat. no. 415–13)
- Sheet of porous membrane (e.g. Costar Corp. Nucleopore, cat. no. 112107 or Millipore filter, cat. no. HATF 142–50)
- Autoclave
- Matrigel (Collaborative Biomedical Products): dilute 1:8 with sterile distilled water

Method

1. Wash the Pyrex tubing cylinders in a solution of 7X detergent, rinse several times with distilled water and dry.

2. Place a small amount of 732 RTV-sealant on a flat surface and spread it out with a spatula to form an even film (~0.2 mm thick).
3. Place each glass cylinder on the film so the entire edge is covered with a *thin* layer of sealant.[b]
4. Place the sealant-covered end of each cylinder on to a sheet of porous membrane and leave them to dry (undisturbed) for 24 h at room temperature.
5. Remove the excess membrane surrounding the cylinders.[c]
6. Using a 1 ml syringe, create three 2 mm high 'legs' on each membrane by placing three small drops of the RTV-sealant 120° apart on to the bottom edge of each filter. Allow at least 24 h for the sealant to dry.
7. As a quality check, place distilled water into each chamber and allow them to stand for 3–5 min. Reprocess any chambers that leak (see step 10).
8. Place properly sealed chambers inside Petri dishes and sterilize by autoclaving.
9. Prior to use, the filters can be coated with extracellular matrix preparation (e.g. Matrigel; see Chapter 7, *Protocol 2*). Nucleopore membranes *must* be coated!
10. After the chambers have been used, cut the filters away from the glass cylinders using a sharp blade, and soak them in concentrated sulfuric acid for 1–2 days. Thoroughly rinse the cylinders with distilled water and allow them to dry. Reprocess the cylinders to form new chambers by repeating steps 1–9.

[a] The cylinder edges must be made perfectly flat and smooth (e.g. by using sandpaper) to ensure that contact with the filter surface is leak-proof, and the cells are evenly distributed when plated.
[b] This is critical as excess sealant may spill on to the filter (see step 4) and prevent cell attachment, while an inadequate amount of sealant will result in leakage.
[c] Nucleopore and Millipore membranes can be removed by rapid flaming and by cutting with a sharp blade, respectively.

2.3 Sertoli cell isolation

The major steps in the Sertoli cell isolation procedure are shown diagrammatically in *Figure 3* and detailed in *Protocol 2*. We have used this procedure successfully for obtaining Sertoli cells from rats ranging from 10 to 60 days of age. However, it is more difficult to obtain a uniform and highly enriched Sertoli cell preparation from older animals due to the greater abundance of germ cells. In general, the procedure needs to be adjusted according to the animal age and species being used. The purity of freshly isolated Sertoli cells

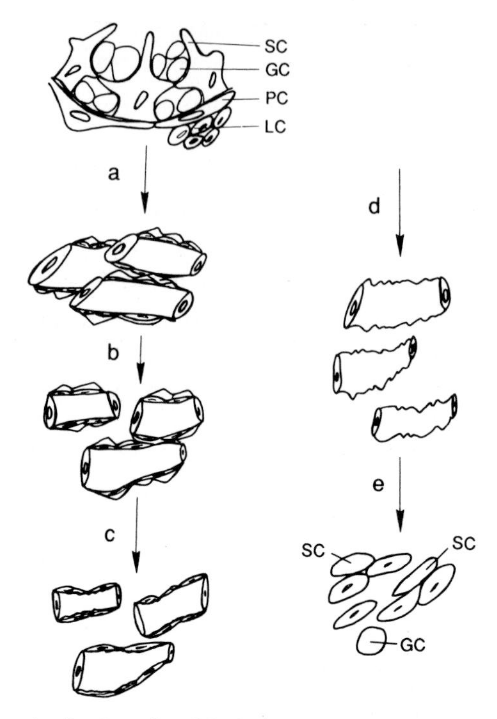

Figure 3. Major steps in the Sertoli cell isolation procedure (see *Protocol 2*). (a) Testicular tissue is minced with sharp blades into 1–2 mm fragments. At this stage the cellular elements include: SC, Sertoli cells; GC, germ cells; PC, peritubular myoid cells; LC, Leydig cells and other. (b) The minced tissue is incubated with a trypsin solution to sepaate the seminiferous tubules from the interstitial tissue. (c) The bulk of the interstitial cells are removed by repeated sedimentation steps. (d) The tubule fragments are incubated with collagenase to remove the peritubular myoid cells. (e) Following several sedimentation steps to eliminate most of the peritubular myoid cells and germ cells, the tissue is mechanically dispersed using a Pasteur pipette, and the larger Sertoli cells and Sertoli cell clusters are separated by mild centrifugation. The various stages can be determined by examining a small drop of the preparation under a phase-contrast microscope (reproduced from A. Steinberger and J.-P. Clinton, *Methods in toxicology*, Vol. 3A, 1993, with permission of Academic Press, Inc.).

usually ranges between 80 and 90%, and the Sertoli cells should comprise > 95% of cells in culture after several media changes. The identifying features of Sertoli cells in monolayer cultures on non-porous supports have been documented in detail elsewhere (11, 12) and are illustrated in *Figure 1*. When examined with phase-contrast optics, the Sertoli cells exhibit a pale cytoplasm with irregular nuclei and prominent nucleoli.

Protocol 2. Isolation and culture of Sertoli cells

Equipment and reagents

- 250 ml of Hanks' Balanced Salt Solution without $NaHCO_3$ (HBSS): dilute 25 ml of 10 × concentrated HBSS without $NaHCO_3$ (Gibco–BRL, cat. no. 310–3180) to 250 ml triple-distilled water. Adjust the pH to 7..3–7.4 with NaOH[a]
- 250 ml of Ca^{2+}, Mg^{2+}-free HBSS containing 0.02% (w/v) EDTA: add 47.5 mg of EDTA to 250 ml of Ca^{2+}, Mg^{2+}-free HBSS and adjust the pH to 7.4 using NaOH[a]
- 250 ml of HBSS containing 0.2% (w/v) BSA: dilute 10 × concentrated HBSS without $NaHCO_3$ 1:10 with triple-distilled water. To 250 ml of this solution add 500 mg of BSA. Stir for 25–30 min until the protein is thoroughly dissolved[a]
- 30 ml of 0.25% (w/v) trypsin solution: dilute 10 × concentrated HBSS without $NaHCO_3$ 1:10 with triple-distilled water. Add 75 mg of trypsin (Sigma, cat. no. T-0646) and 140 mg of Hepes buffer (Sigma, cat. no. H-3375) to the solution and stir it for 15–20 min. Adjust the pH to 7.4 with NaOH and sterilize the solution by filtration through a 0.22 μm Millipore filter. Keep the solution frozen at −20 °C until required and use within one month of preparation
- 100 ml collagenase solution: to 90 ml of HBSS, add 0.47 g of Hepes buffer (final concentration, 20 mM) followed by 100 mg of BSA (Sigma, cat. no. A-7906). Stir the solution for 15 min and adjust the pH to 7.4 with NaOH. Add 150 mg of collagenase (Sigma, Type II), 150 mg of hyaluronidase (Sigma, Type I) and 1 ml of 1 mg/ml DNase (Sigma, Type I, cat. no. D-5025) to the solution and stir for 15 min. Adjust the volume to 100 ml with HBSS and correct the pH to 7.4–7.5 with NaOH. Prefilter using a paper filter and sterilize by filtering through a 0.22 μm Millipore filter. Store frozen at −20 °C in 30–35 ml aliquots and use within one month of preparation
- A tissue-mincing device: we use 26 razor blades mounted on a holder 1 mm apart to manually cut the testes into small (1 mm) pieces
- Thumb tacks
- 20 ml syringe (Beckton Dickinson, cat. no. 3661)
- Matrigel coated filters in 12-well plates (see *Protocol 1*)
- 2 litres DMEM/Ham's F-12 (1:1): add and thoroughly dissolve 1 packet of Ham's F-12 (ICN–Flow, cat. no. 10–421–20), 1 packet of DMEM (ICN–Flow, cat. no. 10–331–20) and 7.3 g of Hepes buffer (Sigma, cat. no. 3375) to 1.8 litres of distilled water. Add 2.4 g of $NaHCO_3$ (Fisher, cat. no. S-233) and stir for another 10 min. If necessary, increase the pH to 7.2 by adding 1 M NaOH. Adjust the total volume to 2 litres and sterilize by filtering through a 0.2 μm AcroCap filter (Gelman Sci., cat. no. 4480). Store as 500 ml volumes at 4 °C[c]
- Culture medium: *on the day of use*, add 250 μl of 5 mg/ml gentamicin (Sigma, cat. no. G-1014), 5 ml of fungizone (Gibco–BRL, cat. no. 15295; diluted as indicated on the lyophilized preparation), 5 ml of ITS (insulin, transferrin, sodium selenite; Collaborative Biomedical Products, cat. no. 40351 or ITS+, cat. no. 40352), 5 μg of EGF (Collaborative Biomedical Products, cat. no. 40001[b]), 50 μl of 2 mg/ml Vitamin A (Sigma, cat. no. R-0635; dissolved in water) and 50 μl of 2 mg/ml Vitamin E (Sigma, cat. no. T-3634; dissolved in ethanol) to 500 ml of DMEM: Ham's F-12
- 1 mg/ml DNase solution (Sigma, Type I, cat. no. D-5025)
- Large, medium, and small scissors
- Large, medium, and small forceps (stainless steel instruments sterilized prior to use)
- A container for killing the animals
- CO_2 gas (house supply or a tank of compressed CO_2)
- 70% (v/v) ethanol
- Large (100 mm) Petri dish
- Surgical table (e.g. a cork plate covered with absorbent paper)
- Clinical centrifuge (IEC-Model CL)
- 250 ml Erlenmeyer flask[d]
- Shaking waterbath

A. *Removal of the testes*

1. Kill the animal with an overdose of CO_2 and place it on its back on the surgical table. Spread its legs apart and pin them to the table using thumb tacks.
2. Soak the animal's abdomen with 70% ethanol in order to sterilize the

Protocol 2. *Continued*

area to be operated on and to prevent dry hair from falling into the animal's abdominal cavity when it is opened.

3. Lift the skin with large forceps (to avoid perforating the abdominal muscle) and, using large scissors, cut the skin across and down toward the hind legs.
4. Pull the skin flap down to expose the abdominal muscle. Steps 1–4 can be performed on several animals before proceeding to step 5.
5. From this point onwards, all the materials used must be sterile and hands should be gloved. Lift the abdominal muscle with medium-sized forceps and, using medium-sized scissors, make a 2 cm vertical incision about 3 cm above the pubis.
6. Insert small forceps through the incision and pull up the testis about 1 cm above the incision. Cut the testis free and place it in a sterile dish containing sufficient Ca^{2+}, Mg^{2+}-free HBSS/EDTA solution to submerge the tissue. Repeat this step for the other testis and steps 5 and 6 for the other animals. Remove the dish to a laminar-flow hood.

B. *Decapsulation and initial dispersion of the tissue*

1. Grasp each testis at the upper pole with forceps, cut open the tunica albuginea with fine scissors, and separate the testicular tissue from its capsule.
2. Transfer the testicular tissue to the cork plate and cut it in four different directions with the mincing device. Transfer the minced tissue into a 25 ml beaker[d] containing 10–15 ml Ca^{2+}, Mg^{2+}-free HBSS/EDTA solution.[e]
3. Using a 20 ml syringe without a needle, disperse the tissue by aspirating the contents 5–8 times. Transfer the contents into a 50 ml centrifuge tube and allow 3–4 min for the tissue to sediment.
4. Using a sterile Pasteur pipette connected to vacuum line, aspirate and discard the supernatant. Resuspend the tissue in 50 ml of HBSS and allow another 3 min of settling time.

C. *Trypsin digestion*

1. After discarding the supernatant, add warm (36.5 °C) 0.25% trypsin solution (30 ml for 16–24 testes) and transfer the contents to a 250 ml flask containing 1 ml of 1 mg/ml DNase-1 solution.
2. Place the flask in a waterbath set at 36.5 °C and oscillating at 80 cycles/min. Usually about 30–45 min are needed to fully disperse the seminiferous tubules, however it is good practice to intermittently check the degree of dissociation visually as the proteolytic activity of trypsin preparations can vary significantly.

3. Continue the incubation until the tubules appear adequately dispersed (see *Figure 3*). If clumping of the tissue occurs, add a further 0.5–1 ml of the DNase solution.

D. *Collagenase digestion*

1. Following the trypsin digestion, carefully decant the dispersed tissue into a 50 ml centrifuge tube and allow the tubules to sediment for 3–4 min.
2. Aspirate the supernatant (predominantly interstitial cells) and wash the residual tubules with 40 ml of HBSS/0.2% BSA solution. Leave the suspension to sediment for 3–4 min.
3. Add 30 ml of collagenase solution to the sedimented tissue and transfer it immediately into a 250 ml Erlenmeyer flask.
4. Place the flask back into the waterbath.[f]
5. Following the incubation, transfer the tissue contents to a 50 ml centrifuge tube and allow them to sediment under gravity for approximately 8 min.
6. Aspirate most of the supernatant (mainly peritubular myoid cells) and wash the residue with HBSS/0.2% BSA.
7. Repeat step 6 until the supernatant appears clear. Discard the supernatant to 0.5–1 cm above the tissue.
8. Using a sterile, siliconized, and cotton-plugged Pasteur pipette with a fire-polished 0.5 mm diameter tip, disperse the sedimented Sertoli cells by repeated aspiration.
9. Add 25 ml of HBSS/0.2% BSA to the cell suspension and let the contents settle for approximately 5 min.
10. Transfer the supernatant to another centrifuge tube and repeat steps 8 and 9 for the remaining sediment. Combine the second supernatant with the first.[g]
11. Centrifuge the cell suspension in a table-top clinical centrifuge at the *lowest speed* for approximately 3 min to sediment the cell aggregates but not the single cells.
12. Remove the supernatant as completely as possible without disturbing the sediment and discard it. Resuspend the sedimented cells in 10 ml of HBSS/0.2% BSA.
13. Place a drop of the cell suspension on a glass slide and check its quality using a phase-contrast microscope. A successful isolation should yield ≥ 80% Sertoli cells singly or in small (≤ 10 cells) aggregates.[h] If necessary, repeat steps 11 and 12 one or more times.

Protocol 2. *Continued*

14. Count the cells in a haemocytometer. On average, the number of Sertoli cells that can be obtained from a single testis of an 18-day-old rat is about 5×10^6 of which > 95% are viable as determined by Trypan Blue exclusion.

E. *Culture of Sertoli cells on filters*

1. If the cell preparation appears satisfactory (see part D, step 14), wash it twice with serum-free culture medium and dilute it appropriately for plating on to filters.
2. Seed 2×10^6 Sertoli cells/1.8 ml culture medium into each filter (i.e. approximately 0.8×10^6 cells/cm^2).[i]
3. Add 1.8 ml of culture medium to each outer chamber. The level in both chambers should be similar for proper hydrostatic pressure.
4. Ensure that no air is trapped under any of the filters.
5. Incubate the plates in a humidified atmosphere of 2% CO_2:98% air.[j] Do not disturb the plates for at least 24 h to ensure that the cells attach firmly.
6. Change the medium at 1–2 day intervals. Most residual germ cells in the dishes are usually removed during the first 2–3 media changes. With regular refeeding, Sertoli cell cultures can be maintained for many weeks.

[a] Sterilize the solution after preparation by passing it through a 0.22 µm Millipore filter, and store it at 4 °C until required.
[b] Dilute a vial containing 100 µg of EGF with 10 ml of distilled water, and store frozen at –20 °C as 500 µl aliquots.
[c] Two–three aliquots (1.0–1.5 ml each) of the DMEM/Ham's F-12 medium should be incubated at 37 °C for at least three days prior to use to check for any contamination.
[d] All glassware coming into contact with the cells must be siliconized by coating the inner surface with a solution of ethyl ether (Fisher, cat. no. E134) containing a drop of antifoam concentrate (Sigma, cat. no. A-5623). After a few seconds contact with the solution, dry the glassware in an exhaust hood (the fumes are toxic!), and sterilize as normal. The siliconized glassware can be reused several times before it needs to be recoated.
[e] We have adapted this volume for 16–24 testes from 18–20-day-old rats. The volume should be adjusted if more or less tissue is used.
[f] Since the time needed for collagenase action will vary depending on the activity of the enzyme, age of donor animals, amount of tissue, etc., it is essential to check the preparation at 15 min intervals following the addition of the collagenase to determine if the tubules are free from peritubular myoid cells. This is best accomplished by placing a tiny drop of the suspension on a slide and examining it with a phase-contrast microscope. A large number of 'smooth' tubules indicates insufficient collagenase action (see *Figure 3*) and the need for a longer incubation.
[g] The goal is to obtain small aggregates of ≤ 10 Sertoli cells. Insufficient or excessive digestion will adversely affect the uniformity and viability of the cultures.
[h] Sertoli cells are the largest cells in the testis and are easily recognized by their pale cytoplasm and large, irregular nucleus. In rodents the nucleus has a characteristic 'tripartite nucleolus'.
[i] A high plating density is required for the formation of a confluent columnar monolayer of Sertoli cells.
[j] The cells can be incubated at 33 or 36.5 °C.

2.4 Testing for monolayer and tight junction patency

2.4.1 Permeability of monolayers to [^{3}H]inulin

The integrity of epithelial cell monolayers on porous membranes can be assessed by measuring the passage of large, hydrophilic compounds through the cell layer (e.g. HRP, see Chapter 3, *Protocol 4*; mannitol, see Chapter 6, *Protocol 2*). To assess the permeability of Sertoli cell monolayers, we add 0.1 μCi of [^{3}H]inulin (specific activity, 260 mCi/μg; ICN Radiochemicals) to the *outer* chamber (i.e. the *basal* compartment), and after 3.5 h measure the radioactivity in the inner (apical) compartment. Results are expressed relative to the values obtained in control (cell-free) chambers (4, 6, 7). The permeability of Sertoli cell monolayers to [^{3}H]inulin is usually reduced by approximately 90% after 3–5 days of culture, and remains unchanged for at least 13 days (see *Figure 4*). The decline in permeability appears to be correlated with the formation of tight junctions (as determined by electron microscopy). Formation of confluent monolayers of peritubular myoid cells, which do not form tight junctions, only slightly reduces [^{3}H]inulin diffusion, providing additional evidence that a decline of [^{3}H]inulin diffusion across Sertoli cell monolayers may, indeed, be related to the formation of tight junctions.

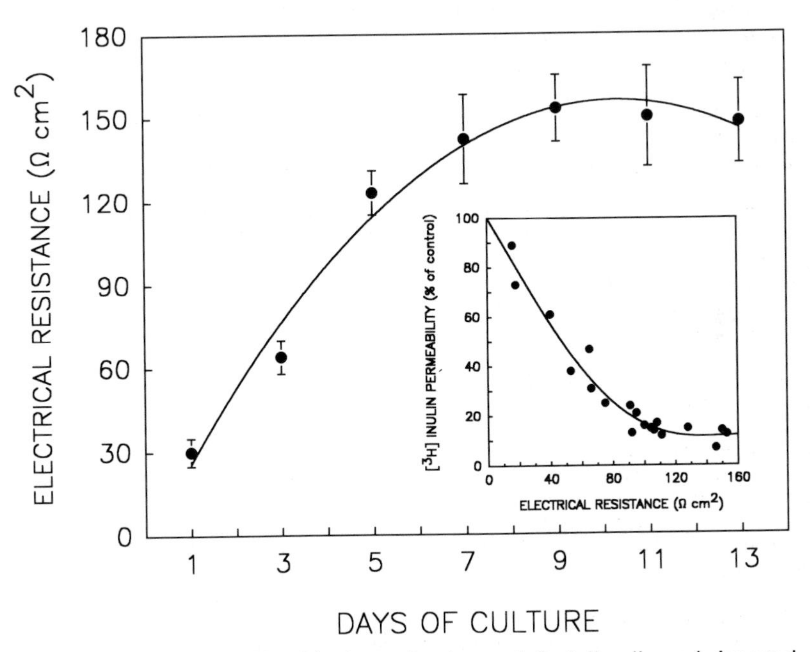

Figure 4. The change in TER with time of culture of Sertoli cells and, inserted, the inverse relationship between [^{3}H]inulin permeability and TER. Note that the TER increases after [^{3}H]inulin permeability has plateaued (reproduced from A. Janecki, A. Jakubowiak, and A. Steinberger, *Endocrinology*, Vol. 127, 1990, with permission of The Endocrine Society).

2.4.2 Transepithelial electrical resistance (TER)

The [^{3}H]inulin permeability test can only be applied to a culture once due to radioactive contamination and is consequently unsuitable for continual monitoring of monolayer patency. Measuring TER, for example using the ERS device (Millipore) eliminates this problem and is a convenient and sensitive indicator of tight junction status (refs 7–9; see also Chapter 6, Section 2.3.1).

To avoid temperature-dependent fluctuation of TER, the cultures should be stabilized for 25 min at 27 ± 1 °C prior to measurements of resistance. We have found that the TER values at 27 °C are approximately 12% and 16% higher than those obtained at 33 and 36.5 °C, respectively, and remain stable for at least 2 h. The final values of TER are calculated by subtracting the mean (n = 3) resistance of cell-free, Matrigel-covered culture chambers from the mean (n = 3) resistance of cell monolayer plus support, and multiplying the difference by the surface area of the filter support. The results are expressed as ohms.cm^2 (see *Figure 4*).

The main advantage of the TER method is that it can be used repeatedly during an experiment without affecting cell viability. Moreover, since resistance of most epithelia seems to be related to the complexity of tight junctions (13, 14), the TER values reflect the status of tight junctions *in vitro*. The results of our earlier studies clearly demonstrated that the formation of tight junctions is accompanied by a dramatic increase of TER, and is hormone- and temperature-dependent (7, 8).

Although all of the above procedures have been used successfully in our and several other laboratories, they do require a little practice and individual judgement. Cell culture is still to a large degree an art! Good luck!!

3. Effects of testicular toxicants on the blood–testis barrier *in vitro*

3.1 Cadmium chloride

We have utilized two-compartment cultures for exploring the effects of cadmium chloride ($CdCl_2$) on several Sertoli cell functions, including cell viability, vectorial secretion of inhibin, and the development of intercellular tight junctions (9). These studies were based on our previous findings that Sertoli cell monolayers cultured for several days in the two-compartment culture system, and in the presence of FSH and testosterone, developed high TER values (7).

Toxic effects of $CdCl_2$ on testicular morphology and function *in vivo* have been reported by several investigators (for review, see ref. 11), however, the mechanism of cadmium toxicity in the testis is unknown. Vascular changes, interstitial oedema, decreased capillary blood flow, and subsequent ischaemia have all been suggested as major contributory factors (15). On the other hand, low doses of cadmium were reported to compromise spermatogenesis without noticeable changes of the vasculature (16), suggesting that cells other than endothelial cells may be affected by the toxicant. Setchell and Waites (17) observed a rapid increase in the permeability of the 'blood–testis' barrier following a single injection of $CdCl_2$ suggesting an acute effect on inter-Sertoli cell-tight junctions. Also, data published by Johnson (18) suggested a cadmium effect on the 'blood–testis' barrier in guinea-pigs.

We investigated the effects of $CdCl_2$ on the development of intercellular tight junctions in two-compartment cultures of immature (18-day-old) rat Sertoli cells (9). The cultures were prepared as described in *Protocol 2* except that sodium selenite was omitted from the medium. The cells were exposed to various doses of $CdCl_2$ (0.75–24 μM) for 4–18 h on days 1 or 5. After exposure, the cultures were washed and reincubated in control medium without $CdCl_2$ for an additional 8 days. The status of the tight junctions was monitored by repeated measurements of TER (usually every 48 h). For defining the site of $CdCl_2$ effects, the TER changes were correlated with Sertoli cell number (DNA content, see *Protocol 3*), viability (MTT test, see *Protocol 4*), and secretory activity (immunoreactive inhibin, see *Protocol 5*) (9).

Protocol 3. Assessment of cell number by a modification of the method of Labarca and Paigen (19)[a]

Equipment and reagents

- Assay buffer: 0.05 M sodium phosphate buffer/2 M NaCl, pH 7.4
- bisBenzimide (Hoechst, cat. no. 33258) dye solution (3 μg/ml)
- Sonicator
- Spectrofluorometer
- Calf thymus DNA standard preparation (Sigma)

Method

1. Wash the cells twice with assay buffer and rupture them by sonication. Sonicate in 1 ml of assay buffer using short (10 s) bursts at maximum power until the solution clears.
2. Centrifuge at 12 000 *g* for 10 min at room temperature to remove the debris.
3. Sonicate the supernatant again and dilute 1:10 with the assay buffer. If the homogenates are not used immediately, 2 mM EDTA should be added to prevent DNase activity.
4. Add 1 ml aliquots of each sample to 2 ml of bisBenzimide dye solution (final dye concentration, 1 μg/ml).
5. Incubate for 10 min at room temperature in the dark.
6. Determine the fluorescence in a spectrofluorometer with excitation emission wavelength set for 360 nm and 460 nm, respectively.[b]

Protocol 3. *Continued*

7. Express the data relative to a DNA standard preparation (0.28–36 μg DNA/tube).

[a] Since Sertoli cells from 18-day-old rats do not divide *in vivo* or *in vitro*, DNA content provides a good estimation of cell number.
[b] The fluorescence measurements can be made at any time after adding the dye provided the samples are kept away from strong direct light.

Protocol 4. Assessment of cell viability using the MTT assay of Hefti and Hauck (20)[a]

Equipment and reagents

- 5 mg/ml MTT (Sigma, cat. no. M-2128) stock solution prepared in phosphate buffer and filtered through a 0.22 μm filter. For use, dilute the stock solution 1:10 with culture medium (see *Protocol 2*)
- HBSS (see *Protocol 2*)
- 5% (w/v) SDS in water
- Spectrophotometer

Method

1. Aspirate the medium and wash the cells once.
2. Add 0.5 ml of the MTT solution (0.5 mg/ml) to the cells and leave them at 37 °C for 4 h.
3. Wash the cells three times with HBSS.
4. Add 1 ml of 5% SDS solution to solubilize the cells. Sonicate for 10 s at maximum power and incubate at 37 °C for 24 h.
5. Read the absorbance on a spectrophotometer set at 550 nm; dilute 1: 1 and 1:3 in water.

[a] The MTT assay is based on the reduction of yellow tetrazolium salt (MTT) to blue formazan by the mitochondrial succinate dehydrogenase enzyme complex. The amount of formazan dye is directly proportional to the cell number and decreases rapidly in non-viable cells. Thus, changes in the amount of dye formed represent changes in cell number and/or viability.

Protocol 5. Inhibin radioimmunoassay as described by Janecki *et al.* (6)

Equipment and reagents

- Iodination buffer: 1.7 g NaH_2PO_4 and 1.8 g Na_2HPO_4 in 250 ml of water (pH 7.4)
- Assay buffer: 7.1 g NaH_2PO_4, 6.0 g EDTA–Na_2, 2.0 g sodium azide, 1 ml normal rabbit serum, 1.0 g BSA, and 1.0 g Triton X-100 in 1 litre of water (pH 7.4)
- [^{125}I]NaI (1.0 mCi/10 μl; ICN Biochemicals)
- Synthetic porcine N-terminal porcine α-inhibin (1–20) provided by Dr N. Ling (Salk Institute)
- PEG solution: 8% PEG_{6000} in 0.9% (w/v) NaCl

- Antibody to the porcine α-inhibin (1–30) provided by Dr S.-Y. Ying (Salk Institute). Final dilution 1:250 000. This antibody did not cross-react with synthetic GnRH, FSH, TGFβ, LH, TSH, PRL, GH, relaxin, or activin
- Secondary antibody: anti-rabbit gamma globulins (Antibodies Inc.), 1:20 in assay buffer with 1% normal rabbit serum
- Assay tubes: borosilicate 12 × 75 mm culture tubes
- Gamma counter
- Iodogen-coated tubes (iodogen film of 0.1 mg/ml of iodogen solution in dichloromethane)
- Sephadex G-50 (Pharmacia) column (19 × 1 cm) eluted with assay buffer

A. *Iodination of the α-inhibin 1–30 fragment according to the method of Salacinski* et al. *(21)*

1. Wash the iodogen-coated tube with 100 μl of iodination buffer.
2. Add 10 μl of iodination buffer.
3. Add 10 μl (5 μg) of α-inhibin.
4. Add 10 μl (1.0 mCi) of [^{125}I]NaI.
5. Mix and incubate for 10 min at room temperature.
6. Add 200 μl of iodination buffer and incubate for an additional 5 min.
7. Load on to the column and elute with assay buffer.
8. Collect fractions; 15 drops/tube containing 1.0 ml of assay buffer.
9. Count 10 μl aliquots to determine the position of the iodinated inhibin (with a specific activity of approximately 100 μCi/μg).
10. Aliquot and store the iodinated inhibin at –20 °C for up to 1 month until use.

B. *Radioimmunoassay of iodinated inhibin*[a]

1. In duplicate, dispense the standards and unknown samples into a total volume of 400 μl/tube using assay buffer as the diluent. Use the same volume of assay buffer instead of sample/standards for blanks and reference tubes.
2. Add 200 μl of primary antibody to the reference, standard, and unknown tubes. To the blank tubes add 200 μl of assay buffer.
3. Mix and set aside for 24 h at room temperature.
4. Add 100 μl of iodinated inhibin (approximately 20 000 c.p.m./tube).
5. Mix and set aside for 24 h at room temperature.
6. Add 200 μl of secondary antibody solution (1:20) and shake.
7. Add 1 ml of PEG solution and shake.
8. Incubate for 1 h at room temperature.
9. Centrifuge at 2000 *g* for 1 h at 4 °C.
10. Decant the supernatant and count in a gamma counter.
11. Use four-parameter logistic curve fitting to calculate doses.

[a] The usable range of this assay was 0.8–60 fmol/tube (80–20% of binding) and the limit of sensitivity was 0.1 fmol/tube. The inter- and intra-assay variations were 4.5% and 9%, respectively.

The results of our experiments (see *Figure 5*) showed that the effects of $CdCl_2$ *in vitro* depend on the concentration, as well as the onset and duration of exposure of the Sertoli cells to the toxicant. The observed effects could be divided into four principal categories:

(a) At highest *cumulative doses* (concentration × duration of exposure), the TER values decreased significantly and irreversibly during 13 days of culture, the decrease being accompanied by a significant and irreversible drop in inhibin secretion, cell viability, and cell number.

(b) Within a narrow range of doses, an irreversible, or partly reversible, decline of TER was accompanied by a transient decrease, or no change, of secretory activity without significant changes in Sertoli cell number and/or viability.

(c) With lower doses, the TER values rapidly decreased and then returned to the control level within 3–4 days. In this group, no change in either inhibin secretion or cell viability was observed.

(d) Exposure to the lowest doses of $CdCl_2$ caused a delayed, but significant increase of TER. This increase was not accompanied by noticeable changes in other parameters evaluated, suggesting that $CdCl_2$ may selectively affect Sertoli cell tight junctions without affecting cell secretory activity, cell number, or cell viability. However, increasing cumulative doses of $CdCl_2$ led to decreased inhibin secretion and cell viability, and finally to irreversible cell damage and death.

The mechanism by which $CdCl_2$ can selectively affect the tight junctions requires further investigation. Our data suggest that damage of the 'blood–testis' barrier may be one of the earliest events occurring after exposure of the testis to sublethal doses of cadmium.

3.2 Phthalate esters

The culture conditions employed to study the effects of phthalates were similar to those described for $CdCl_2$, however, the phthalate compounds were continuously present in the Sertoli cell cultures for up to 11 days and were replenished at 48 h intervals during medium changes (8). At different time intervals, the effects of mono-2-ethylhexyl phthalate (MEHP; 100 μM and 500 μM) and monoethyl phthalate (MEP; 100 μM) were examined on the same four parameters as described for $CdCl_2$. The results of TER changes are shown in *Figure 6*. MEHP has been reported to exert toxic effects on germ cell attachment to Sertoli cell monolayers in culture (22, 23), and to decrease FSH-stimulated accumulation of cAMP (24) in the Sertoli cells. MEP served as a negative control.

In our study, the 100 μM concentration of MEHP did not significantly affect the TER during the culture period. However, 500 μM MEHP caused a significant and progressive decrease of TER to approximately 20% of the

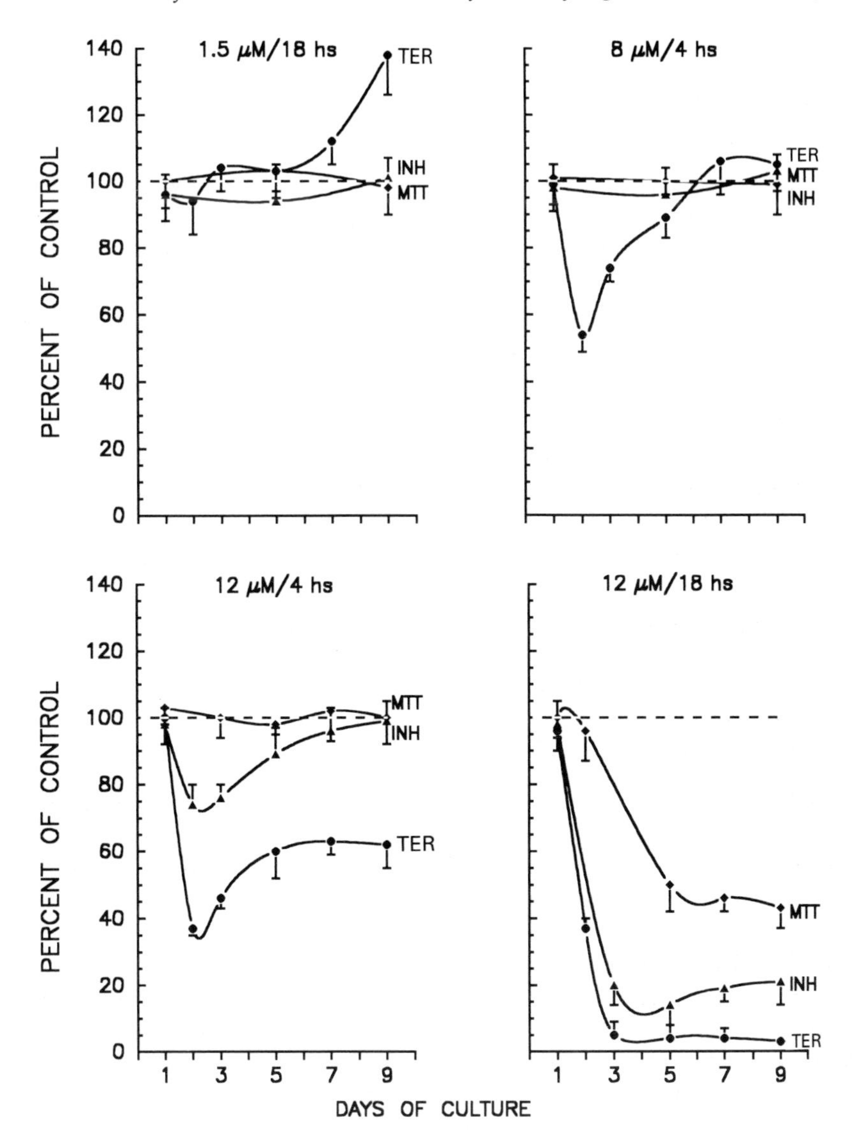

Figure 5. The patterns of transepithelial electric resistance (TER), inhibin secretion (INH), and cell viability (MTT) in Sertoli cell cultures exposed to increasing cumulative doses of $CdCl_2$ (concentration and hours of exposure). $CdCl_2$ was added for 4 h or 18 h on day 1 of culture. The respective $CdCl_2$ concentrations and length of exposure are indicated on each panel. The data are presented as per cent of untreated control values and are means ± SD from triplicate cultures (reproduced from A. Steinberger and G. Klinefelter, *Reproductive toxicology*, Vol. 7, 1993, with permission of Pergamon Press).

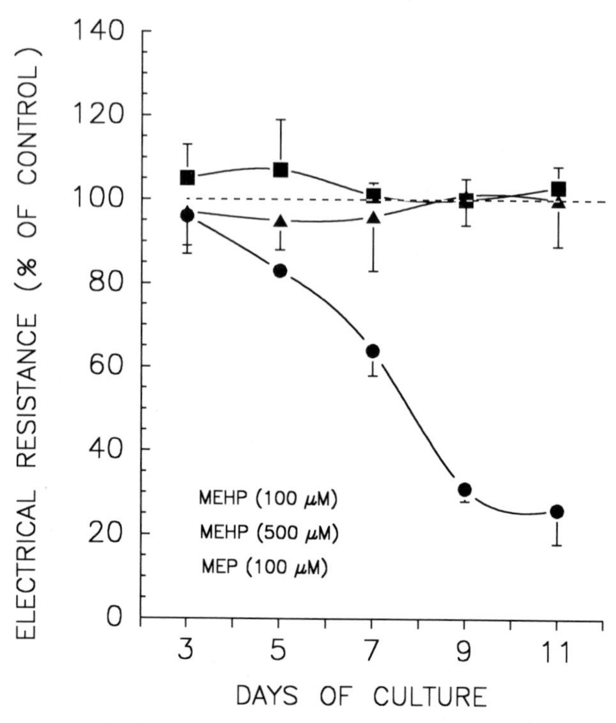

Figure 6. The patterns of TER observed in Sertoli cell monolayers exposed for 11 days to 100 μM (▲) or 500 μM (●) of mono-2-ethylhexyl phthalate (MEHP), or 100 μM of monoethyl phthalate (MEP; ■). Sertoli cell monolayers were incubated at 11 days in the absence (----) or presence of phthalates at the indicated concentrations. Fresh compounds were added at each medium change (every 48 h). The data are means ± SD from triplicate cultures (reproduced from A. Steinberger and G. Klinefelter, *Reproductive toxicology*, Vol. 7, 1993, with permission of Pergamon Press).

initial value after 9 days of culture. Both concentrations of MEHP caused a dramatic decrease of total inhibin secretion after 3 days of exposure, with little change afterwards. On day 11 both the cell number and viability were also decreased significantly in monolayers exposed to 500 μM MEHP. These results indicate that the decrease of TER and inhibin secretion observed in 11-day cultures exposed to 500 μM MEHP was probably due to cell death rather than a specific effect on tight junctions and/or secretory activity. On the other hand, the significant decrease of inhibin secretion noted with the 100 μM concentration of MEHP in the absence of changes in cell viability may have been specific. These results emphasize the importance of measuring multiple-endpoints, including cell number and viability, in order to define the specificity of toxic effects on a particular cell function. The molecular mechanism by which MEHP affects Sertoli cell secretion remains to be elucidated. Data reported by Heindel and Chapin (24) suggest that it may involve suppression of FSH-stimulated cAMP production.

4. Conclusions

In the male, the testis is particularly vulnerable to xenobiotics, yet this organ has been inadequately studied by toxicologists, possibly due to a lack of suitable experimental models. Direct extrapolation of results obtained from Sertoli cell cultures to *in vivo* situations represents a quantum jump and must be done with great caution. However, culture models can provide valuable clues regarding potential sites and mode of action of reproductive toxicants *in vivo*. Moreover, two-compartment cultures can also be useful in exploring the importance of cell–cell interactions in the expression of toxicity on a specific cell type. Chapin *et al.* (25) utilized two-compartment chambers to investigate the effects of tri-*o*-cresyl phosphate in co-cultures of Leydig and Sertoli cells, and showed that this xenobiotic had to be metabolized first by the Leydig cells before exerting a toxic effect on the Sertoli cells.

We believe that two-compartment cultures of Sertoli cells provide a new, powerful tool for investigating the site and mechanism of action of testicular toxicants suspected to cause spermatogenic damage by compromising one or more Sertoli cell functions, including the integrity of the 'blood–testis' barrier. The full potential of this *in vitro* model, however, remains to be explored.

Acknowledgements

Supported in part by NIH grant HD 17802 (AS) and grant 15–149 from the March of Dimes Birth Defect Foundation (AS). The authors are very grateful to Ms Mary Gilliland for typing the manuscript.

References

1. Dixon, R. L. (1984). *Arch. Toxicol.*, **7** (Suppl.), 118.
2. Goldberg A. M. (ed.) (1991). *In vitro toxicology: mechanism and new technology*, Vol. 8. Liebert, NY.
3. Russell, L. D. (1980). *Gamete. Res.*, **3**, 179 (a review).
4. Janecki, A. and Steinberger, A. (1986). *J. Androl.*, **7**, 69.
5. Byers, S. W., Hadley, M. A., and Djakiew, D. (1986). *J. Androl.*, **7**, 59.
6. Janecki, A., Jakubowiak, A., and Steinberger, A. (1990). *Endocrinology*, **127**, 1896.
7. Janecki, A., Jakubowiak, A., and Steinberger, A. (1991). *Endocrinology*, **129**, 1489.
8. Steinberger, A. and Klinefelter, G. (1993). *Reprod. Toxicol.*, **7**, 23.
9. Janecki, A., Jakubowiak, A., and Steinberger, A. (1992). *Toxicol. Appl. Pharmacol.*, **112**, 51.
10. Kelly, C. W., Janecki, A., Steinberger, A., and Russell, L. D. (1991). *Am. J. Anat.*, **192**, 183.
11. Steinberger, A., Heindel, J. J., Lindsey, J. N., Elkington, J. S. H., Sanborn, B. M., and Steinberger, E. (1975). *Endocr. Res. Commun.*, **2**, 261.

12. Steinberger, A., Elkington, J. S. H., Sanborn, B. M., and Steinberger, E. (1975). Culture and FSH responses of Sertoli cells isolated from sexually mature rat testis. In *Hormonal regulation of spermatogenesis* (ed. F. S. French, V. Hansson, E. M. Ritzen, and S. N. Nayfeh), pp. 399–411. Plenum Publishing Co., NY.
13. Gumbiner, B. A. (1987). *Am. J. Physiol.*, **253**, C747.
14. Madara, J. and Hecht, G. (1989). In *Functional epithelial cells in culture* (ed. K. S. Matlin and J. D. Valentich), pp. 131–63. A. R. Liss, NY.
15. Aoki, A. and Hoffer, A. P. (1978). *Biol. Reprod.*, **18**, 579.
16. Lee, I. P. and Dixon, R. L. (1973). *J. Pharmacol Exp. Ther.*, **187**, 641.
17. Setchell, B. P. and Waites, G. M. H. (1970). *J. Endocrinol.*, **47**, 81.
18. Johnson, M. H. (1969). *J. Reprod. Fertil.*, **19**, 551.
19. Labarca, C. and Paigen, K. (1980). *Anal. Biochem.*, **102**, 344.
20. Hefti, A. and Hauck, K. (1990). *In Vitro Cell. Dev. Biol.*, **26**, 63A [Abstract No. 188].
21. Salacinski, P. R. P., McLean, C., Sykes, J. E. C., Clement-Jones, V. V., and Lowry, P. J. (1981). *Anal. Biochem.*, **117**, 136.
22. Gray, T. J. B. and Beamand, J. A. (1984). *Food Chem. Toxicol.*, **22**, 123.
23. Gray, T. J. B. and Beamand, J. A. (1986). *Food Chem. Toxicol.*, **24**, 601.
24. Heindel, J. J. and Chapin, R. E. (1989). *Toxicol. Appl. Pharmacol.*, **97**, 1.
25. Chapin, R. E., Phelps, J. L., Burka, L. T., Abou-Donia, M. B., and Heindel, J. J. (1991). *Toxicol. Appl. Pharmacol.*, **108**, 194.

9

Reconstruction of human skin epidermis *in vitro*

MARIE-CÉCILE LENOIR-VIALE

1. Introduction

Since the pioneering work of Rheinwald and Green in 1975 (1, 2) for establishing keratinocytes in culture, numerous advances have been made in the reconstruction of human epidermis *in vitro*. The more recent techniques are based on the reconstruction of a substrate resembling dermis on which keratinocytes of different sources can be cultivated. After immersion for a few days, these cultures are raised to the air–liquid interface in order to mimic *in vivo* conditions.

In this chapter the most well-established techniques used to reconstitute human epidermis are presented. Two different substrates for culturing keratinocytes on are described: dead de-epidermized dermis (DED), originally developed by Pruniéras *et al.* (3), and the dermal equivalent (collagen–fibroblast lattice) developed by Bell *et al.* (4). The synthetic substrates described by Boyce and Hansbrough (5), Tiollier *et al.* (6), Shahabeddin *et al.* (7), and Rosdy and Clauss (8) are also briefly considered.

Two different sources of human keratinocytes are discussed. The first being of skin biopsies obtained from human volunteers or, more often, from surgical specimens. These skin samples can be dissociated by enzymatic treatment to give single keratinocytes that can be amplified as high-density cultures or by using a feeder layer of growth-arrested 3T3 cells as described by Rheinwald and Green (1, 2). The second source of keratinocytes comes from hair follicles which provide a non-invasive method for obtaining human keratinocytes. Hair follicles can be cultivated on DED or dermal equivalents, as explants or after dissociation.

The reconstruction of epidermis *in vitro* using these methods has been employed to study normal cell differentiation and to reproduce some aspects of pathological conditions (see Section 4). In many cases these models represent suitable alternatives to the use of animals for pharmacological and pharmaceutical research.

2. Techniques used to reconstruct human skin epidermis *in vitro*

2.1 The dermis

2.1.1 DED

Freeman and colleagues (9) were the first to use dead dermis from pig skin as a substrate for human skin explants. To provide the cells with a more physiological substrate, Pruniéras *et al.* (3) developed a method using human DED (see *Protocol 5*). A method for preparing DED is described in *Protocol 1*.

Protocol 1. Preparation of human DED according to Pruniéras *et al.* (3)

Equipment and reagents

- Human skin (e.g. cadaver skin from a skin bank)
- Forceps and sharp scissors
- PBSA (see Chapter 3, *Protocol 1*)

Method

1. Cut the split-thickness human skin into smaller pieces of 2 cm^2.
2. Float the pieces, dermal side down, in PBSA at 37 °C for 7–10 days. Change the PBSA once during this period.
3. Using forceps, separate the epidermis from the dermis.
4. Kill the dermal cells by 10 successive freezings (–20 °C), and thawings.
5. Store the pieces in Petri dishes at –70 °C. The surface from which the epidermis has been removed should be facing upwards.

2.1.2 Dermal equivalent

In 1979, Bell *et al.* (4) proposed the production of a tissue-like structure by incorporating fibroblasts in a collagen lattice. Asselineau and Pruniéras (10) standardized the method of preparation. The protocols for obtaining human dermal fibroblasts from skin (see *Protocol 2*) and for the preparation of dermal equivalents (see *Protocol 3*) are described below.

Protocol 2. Isolation and culture of human dermal fibroblasts according to Régnier *et al.* (11)

Equipment and reagents

- Minimum Essential Medium with Earle's salts and 2.2 g/litre $NaHCO_3$ and without L-glutamine (pH 7.3) (MEM; Gibco–BRL)
- 0.1% (w/v) collagenase (Worthington CLS) in GKN solution
- PBSA (Chapter 3, *Protocol 1*)

- Culture medium: MEM plus 10% (v/v) FCS, 0.1% (v/v) penicillin (10 000 IU/ml)-streptomycin (10000 μg/ml), 1% (v/v) 100 mM sodium pyruvate, 1% (v/v) non-essential amino acids (NEAA), and 1% (v/v) 200 mM L-glutamine (pH 7.3) (all from Gibco–BRL)
- 0.05% (w/v) trypsin–0.02% (w/v) EDTA (Gibco–BRL)
- GKN solution: 0.1% (w/v) glucose, 0.04% (w/v) KCl, and 0.8% (w/v) NaCl. Sterilize this solution by filtration using a filter with a pore size of 0.22 μm
- Sterile gauze
- Curved sharp scissors
- Tissue culture dishes

Method

1. Separate the epidermis from the dermis (see *Protocol 4*, step 7).
2. Cut the dermis into very small pieces ($5mm^2$) using curved scissors.
3. Transfer the pieces of dermis into the collagenase solution and agitate using a magnetic stirrer for 2 h in an incubator at 37 °C.
4. Filter the suspension through the sterile gauze.
5. Centrifuge the filtered suspension for 5 min at 500 *g*.
6. Resuspend the pellet in culture medium.
7. Count the cells using a haemocytometer and seed them into culture dishes at a cell density of 2×10^5 cells/cm^2.
8. Culture the cells in an incubator at 37 °C with an atmosphere containing 5% CO_2. Change the culture medium twice a week.
9. Subculture the cells before they reach confluence by treating them with trypsin–EDTA for 5 min and splitting them 1:3 into new dishes.

Protocol 3. Preparation of dermal equivalents according to Bell *et al.* (4) and Asselineau and Pruniéras (10)

Equipment and reagents

- MEMH: MEM with 25 mM Hepes and without $NaHCO_3$ (pH 7.3)
- Culture medium: MEMH plus 10% (v/v) FCS, 0.1% (v/v) penicillin (10000 IU/ml)–streptomycin (10000 μg/ml), 1% (v/v) 100 mM sodium pyruvate, 1% (v/v) NEAA, 1% (v/v) 200 mM L-glutamine, and 0.125 μg/ml fungizone (amphotericin B) (all from Gibco–BRL). Prepare under sterile conditions
- MEM (1.76 × concentrated): for 100 ml add 17.6 ml MEM (10 × concentrated), 5.16 ml 7.5% (w/v) $NaHCO_3$, 1.76 ml 200 mM L-glutamine, 1.76 ml NEAA, 1.76 ml 100 mM sodium pyruvate, 176 μl 0.1% (v/v) penicillin (10 000 IU/ml)–streptomycin (10 000 μg/ml), 88 μl fungizone, and 176 μl 50 mg/ml gentamicin sulfate to 71.52 ml sterile distilled water. Prepare under sterile conditions
- FCS
- 0.1 M NaOH
- Human fibroblasts: these can be obtained from skin explants (see *Protocol 2*) or as embryonic cell lines from commercial sources, e.g. GM10 cells (ATCC)
- 70% (v/v) ethanol[a]
- Young Sprague–Dawley rats (150–175 g)[a]
- Forceps and scissors[a]
- 0.1% (v/v) and 3% (v/v) (0.5 M) acetic acid (prepared using glacial acetic acid)[a, b]
- 3 mg/ml calf skin collagen, type I (unpepsinized) in 0.5 M acetic acid (Institut Jacques Boy)[b]
- Dialysis tubing, 32 mm diameter, M_w cut off 6000–8000 (SPECTRAPOR ref. 132655)[b]
- 5 litre Erlenmeyer flasks
- 60 mm bacteriological Petri dishes
- High-speed centrifuge and centrifuge tubes[a]

Protocol 3. *Continued*

A. *Preparation of rat tail collagen*

1. Kill the rats by cervical dislocation and remove the tails. The tails may be stored at –20 °C if not required immediately.
2. Wash the tails with cold tap water and soak in 70% (v/v) ethanol for 20 min.
3. Remove the skin and pull the tendons with forceps, one segment at a time, starting from the thin end. Do not allow the tail to become dry. Put the pulled tendons into sterile distilled water and weigh them.
4. Mince the tendons into small pieces with scissors and place the pieces into 0.1% (v/v) acetic acid (final concentration, 1 g tendon/40 ml acid). Keep the mixture at 4 °C for 2 days, stirring occasionally by hand.
5. Centrifuge the mixture at 4 °C for 60 min at 10 000 *g*.
6. After centrifugation, recover the supernatant. This should be clear, ready to use, and contain approximately 2–3 mg collagen/ml.

B. *Dialysis of commercially supplied collagen solutions*

1. Boil the dialysis tubing in distilled water.
2. Dialyse the collagen solution in the dialysis tubing for 48 h at 4°C against 5 litres 0.5 M acetic acid.
3. Continue dialysis in 3 successive baths of 0.1% (v/v) acetic acid for 12 h/bath. Perform this in 5 litre Erlenmeyer flasks, at 4 °C, with magnetic stirring, and under sterile conditions.

C. *Preparation of dermal equivalents*

1. Calculate the volume of each solution required according to the number of dermal equivalents to be prepared. For a one 7 ml volume lattice you will require 3.22 ml MEM (1.76 × concentrated), 0.63 ml FCS, 0.35 ml 0.1 M NaOH, 0.60 ml 0.1% (v/v) acetic acid, and 0.7 ml culture medium containing 2×10^5 fibroblasts. The volume of the lattices can be increased according to the final size and thickness of the dermal equivalents desired.
2. Pool the solutions into a large Erlenmeyer flask containing a magnetic stirring bar. Add the cells last while the solution is gently stirred.
3. Aliquot the pooled solutions into 25 ml Erlenmeyer flasks (5.5 ml/Erlenmeyer).
4. Add 1.5 ml collagen solution slowly and carefully along the inner surface of the flask. The volume of collagen depends on the concentration of the solution. If the collagen solution is too diluted, increase the

volume of collagen solution to be added and decrease the volume of 0.1% (v/v) acetic acid in order to keep the final volume at 7 ml.

5. Mix the pooled solutions vigorously in the 25 ml Erlenmeyer flask (rotary shaking by hand). Do one flask at a time.
6. Pour each pooled solution (7 ml) into a 60 mm bacteriological Petri dish and put it immediately into an incubator at 37 °C and with an atmosphere of 5% CO_2.
7. Gelling occurs within 10–15 min and the gel contracts within a few hours. Check that as the gel contracts it detaches from the edges and the bottom of the dish.

[a] Materials required for preparing collagen from rat tails. One tail contains approximately 1 g of tendons.
[b] Materials required for dialysing commercially supplied collagen.

2.1.3 Synthetic substrates

Synthetic substrates have also been used for reconstituting epidermis, for example lyophilized collagen–glycosaminoglycan (GAG) membranes cross-linked by chemical agents such as glutaraldehyde (5, 12, 13), or chitosan extracted from shrimp shell (7). Dermal substitutes have also be made by cross-linking collagen types I and III and covering this with a film of collagen type IV (6). Finally, inert filters have been shown to be a good substrate for keratinocytes in a chemically defined medium (8).

2.2 The epidermis

2.2.1 Skin biopsies and dissociated keratinocytes

Punch biopsies can be taken (with ethical permission) either from normal volunteers after local anaesthesia, or from discarded skin after surgical procedures, e.g. mammoplasties and circumcisions. Skin obtained from surgical procedures is normally dissociated by enzymatic treatment and the keratinocytes cultivated. This gives rise to a primary culture of keratinocytes which can be amplified by the 3T3 co-culture method originally described by Rheinwald and Green (1, 2) (see *Protocol 4*; *Figure 1*).

Protocol 4. Isolation and culture of human keratinocytes according to Pruniéras *et al.* (3) and Rheinwald and Green (1, 2)

Equipment and reagents

- Human skin explant
- Washing solution 1: 500 ml MEMH (see *Protocol 3*) plus 20 ml antibiotic/antimycotic (ABAM) solution (Gibco-BRL, cat. no. 15240-021) and 160 mg gentamicin sulfate
- Washing solution 2: 500 ml MEMH plus 10 ml ABAM and 80 mg gentamicin sulfate
- Washing solution 3: 500 ml MEMH plus 5 ml ABAM
- PBS (pH 7.3)

Protocol 4. *Continued*

- PBSA (pH 7.3)
- 0.25% (w/v) trypsin in PBSA
- 0.05% (w/v) trypsin–0.02% (w/v) EDTA solution
- FCS; batch tested (14)
- Dulbecco's Modified Eagle's Medium (DMEM) with 4.5 g/litre glucose and 3.7 g/litre $NaHCO_3$, and without L-glutamine (Gibco–BRL)
- HC stock (500 μg/ml): dissolve 5 mg hydrocortisone (Sigma, cat. no. H-4001) in 1 ml 95% ethanol. Dilute 1:10 in culture medium not containing FCS. Store as 0.4 ml aliquots at –20 °C until required
- CT stock: 1 mg/ml cholera toxin (Sigma, cat. no. C-3012) in sterile distilled water. Store at 4 °C
- EGF stock: 10 μg/ml EGF in sterile PBS. Store as 0.5 ml aliquots at –20 °C
- AD stock: 90 mM adenine (Sigma, cat. no. A-9795) in 0.1 M HCl. Store as 1 ml aliquots at – 20 °C
- IN stock: 5 mg/ml insulin (Sigma, cat. no. I-6634) in 0.05 M HCl. Store at 4 °C
- TF stock: 5 mg/ml human transferrin (Sigma, cat. no. T-1147) in sterile PBS. Store 0.5 ml aliquots at –20 °C
- TI stock (2 μM): dilute 6.8 mg tri-iodothyronine (Sigma, cat. no. T-6397) in a minimum volume of 0.02 M NaOH, and make up to 5 ml with ethanol. Dilute 1:1000 in ethanol. Store 0.5 ml aliquots at –20 °C
- Keratinocyte culture medium 1: MEM (see *Protocol 2*) plus 10% FCS, 0.1% (v/v) penicillin (10 000 IU/ml)–streptomycin (10 000 μg/ml), 1% (v/v) 100 mM sodium pyruvate, 1% (v/v) NEAA, and 1% (v/v) 200 mM L-glutamine (pH 7.3) (all from Gibco–BRL)[a]
- 0.02% (w/v) EDTA–0.1% (w/v) glucose in PBSA
- 70% ethanol
- Keratinocyte culture medium 2: as for keratinocyte culture medium 1 plus 0.4 μg/ml hydrocortisone (400 μl HC stock/500 ml medium), 0.1 μg/ml cholera toxin (50 μl CT stock/500 ml medium), and 10 ng/ml EGF (500 μl EGF stock/500 ml medium) (pH 7.3)[a]
- Keratinocyte culture medium 3 (15): as for keratinocyte culture medium 2 except that MEM is replaced by DMEM:Ham's F-12 (3:1) medium and the following additional factors are added; 180 μM adenine (1 ml AD stock/500 ml medium), 5 μg/ml insulin (0.5 ml IN stock/500 ml medium), 5 μg/ml transferrin (0.5 ml TF stock/500 ml medium) and 2 nM tri-iodothyronine (0.5 ml TI stock/500 ml medium) (pH 7.3)[a]
- Mitomycin C solutions: prepare a stock solution of 0.5 mg/ml mitomycin C (Fluka, cat. no. 60824) in sterile PBS. Store in the dark at 4 °C. For use, dilute the stock solution 1:50 in DMEM. Filter sterilize this solution before use
- 3T3 culture medium: as for keratinocyte culture medium 1 except replace MEM with DMEM
- Mouse embryonic 3T3 fibroblasts (Flow; ATCC CCL 92)
- 50 ml sterile centrifuge tubes
- Sterile gauze
- Bench-top centrifuge, 50 ml centrifuge tubes
- Sterile scalpel and two pairs of forceps
- Castroviejo keratome (25 mm cut width)
- Vortex
- Cork board
- Sterile gloves
- Tissue culture dishes
- Filters (pore size, 0.22 μm)

A. *Isolation and high-density culture of keratinocytes*

1. Place the human skin explant in washing solution 2 as soon after removal as possible. Store the tissue at 4 °C and use within three days.
2. Sterilize the cork board with 70% ethanol. Wearing sterile gloves, put the skin explant on to sterile gauze and remove the subcutaneous fat using a scalpel and forceps. Once the fat has been removed, lay the explant on to the cork board.
3. Cut the skin into strips 1 cm wide using a scalpel.
4. While holding the skin with forceps, use a keratome to remove 0.5 mm thick sections from the strips of skin. The blade of the keratome should be at an angle of 45° to the surface of the skin.[b]

5. Wash the strips in washing solutions 1, 2, and 3 for 5 min each time and then wipe them on a piece of sterile gauze.
6. Place the strips, epidermis side up, in 0.25% trypsin and either leave them overnight at 4 °C or for 1 h at 37 °C.[c]
7. Place the strips in trypsin–EDTA solution and separate the epidermis from the dermis using two pairs of forceps. Discard the dermis unless it is required for preparing cultures of fibroblasts (see *Protocol 2*).
8. Transfer the epidermis and trypsin–EDTA solution into a 50 ml centrifuge tube and vortex gently for 2 min. Filter the suspension through sterile gauze and add the same volume of FCS in order to inhibit the action of the trypsin.
9. Centrifuge the suspension for 5 min at 500 *g*. Discard the supernatant and resuspend the cell pellet in culture medium.[a] Count the cells and determine their viability (e.g. using Trypan Blue exclusion). Viability should be > 95%.
10. Seed the cells on tissue culture dishes either at a high density (2×10^5 cells/cm^2) in culture medium 1, or at a lower cell density alongside growth arrested 3T3 cells (see parts B and C). Approximately 40% of the epidermal cells attach. The cells begin to spread out on the dish within 24–48 h.
11. Change the culture medium daily. The dishes will be confluent by 10–15 days.

B. *Culture and cryopreservation of 3T3 fibroblasts*

1. Seed the 3T3 cells into tissue culture dishes at a density of 2000 cells/cm^2. Culture the cells in 3T3 culture medium at 37 °C and in an atmosphere of 5% CO_2.
2. Passage the cells before they reach confluence. Detach the cells by adding trypsin–EDTA solution for 3 min at 37 °C. Split the suspended cells 1:3 into new dishes. If the cells are growing well this should be done twice a week.
3. 3T3 cells should be cryopreserved in 3T3 culture medium plus 10% FCS and 10% DMSO. Use a cell density of 2×10^6 cells/ml.

C. *Co-culture of human keratinocytes and mouse 3T3 fibroblasts (see* Figure 1*)*

1. Rinse subconfluent cultures of 3T3 cells with PBS.
2. Incubate the cells with 10 μg/ml mitomycin C for 2 h at 37 °C.
3. Discard the mitomycin C solution and rinse the cells once with culture medium and once with PBSA.

Protocol 4. *Continued*

4. Treat the cells with trypsin–EDTA for 3 min at 37 °C to detach the cells. Neutralize the trypsin by adding 3T3 culture medium (containing FCS).
5. Seed the 3T3 cells into tissue culture dishes at 1.5×10^4 cells/cm^2. Growth-arrested 3T3 cells can be stored in liquid nitrogen until required.
6. Seed the keratinocytes (see part A) at a cell density of 4000 cells/cm^2 either with the 3T3 cells, or after the fibroblasts have attached (2 h–2 days). In the latter case, add the keratinocytes to the 3T3 cells without changing the medium.
7. Culture the cells in either keratinocyte culture medium 2 or 3, and change the medium twice a week.
8. The cells should be passaged as follows before they reach confluence (7–10 days of culture):
   (a) Rinse the cells with PBSA.
   (b) Add 0.02% EDTA–0.1% glucose solution to the dishes.
   (c) Observe the detachment of the 3T3 cells for about 10 min.
   (d) Encourage the detachment of the 3T3 cells by rigorous pipetting.
   (e) Rinse the remaining attached cells (mainly keratinocytes) twice with PBSA.
   (f) Treat the keratinocytes with trypsin–EDTA for 5 min at 37 °C, twice if necessary, to remove them all from the dishes.
   (g) Reseed and culture the keratinocytes according to steps 6 and 7.
9. Cryopreserve excess keratinocytes in keratinocyte culture medium plus 10% FCS and 10% DMSO at a cell density of 4×10^6 cells/ml.

[a] For high-density cultures use keratinocyte culture medium 1. For co-cultures with 3T3 fibroblasts use either keratinocyte culture medium 2 or 3.
[b] As an alternative to cutting the skin into strips and using a keratome, the explant can be either minced using scissors or the epidermis can be scored with a scalpel at 1 mm intervals.
[c] 0.25% (w/v) dispase (protease) can be used in place of trypsin.

2.2.2 Hair follicles

Hair follicles can be plucked easily and without pain from the scalp of volunteers without the risk of infection or the need for medically qualified personnel (16). Since the follicle will regrow, it can be considered to be an infinite source of keratinocytes. The hair follicle can be used as an explant, or the associated keratinocytes can be cultured (17).

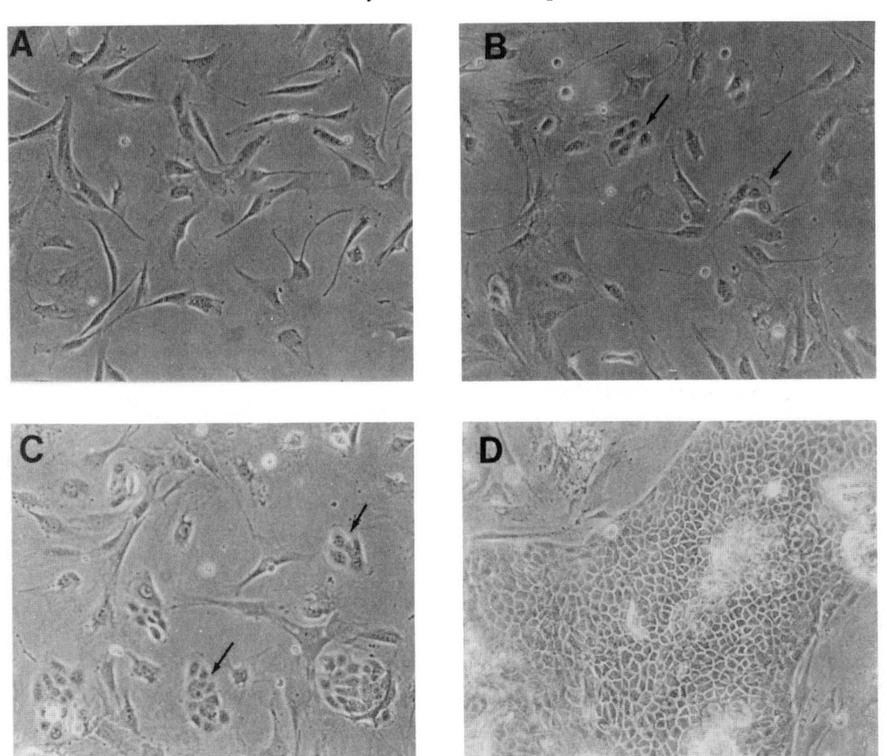

Figure 1. Co-cultures of human keratinocytes and mouse 3T3 fibroblasts. (A) Mitomycin C treated 3T3 cells 24 h after seeding. (B, C) Small epithelial islets (arrows) 24 h (B) and 48 h (C) after seeding keratinocytes on a 3T3 feeder layer. (D) An islet of keratinocytes after one week of culture.

2.3 Reconstruction of human skin epidermis

Keratinocytes obtained by the methods described above (see *Protocol 4*) can be cultivated on either DED or dermal equivalents. *Figure 2* summarizes the steps for reconstituting human epidermis using dissociated keratinocytes and dermal equivalents (see *Protocols 3, 4,* and *6*). Four methods are detailed below (see *Protocols 5–8*) which give rise to reconstructed tissue *in vitro* which resembles normal human epidermis. Other combinations have been described in the literature: punch-biopsies can be inserted into a freshly cast dermal equivalent (19) or deposited on DED (20), and outer root sheath cells can be seeded on lattices (21, 22).

Each method for reconstituting human skin epidermis has its advantages and disadvantages. The preparation of collagen–fibroblast lattices is time-consuming, requires the maintenance of dermal fibroblasts in culture, and is dependent upon a source of good quality collagen. Moreover, lattices are not mechanically resistant and are often subject to collagenolysis (23). For

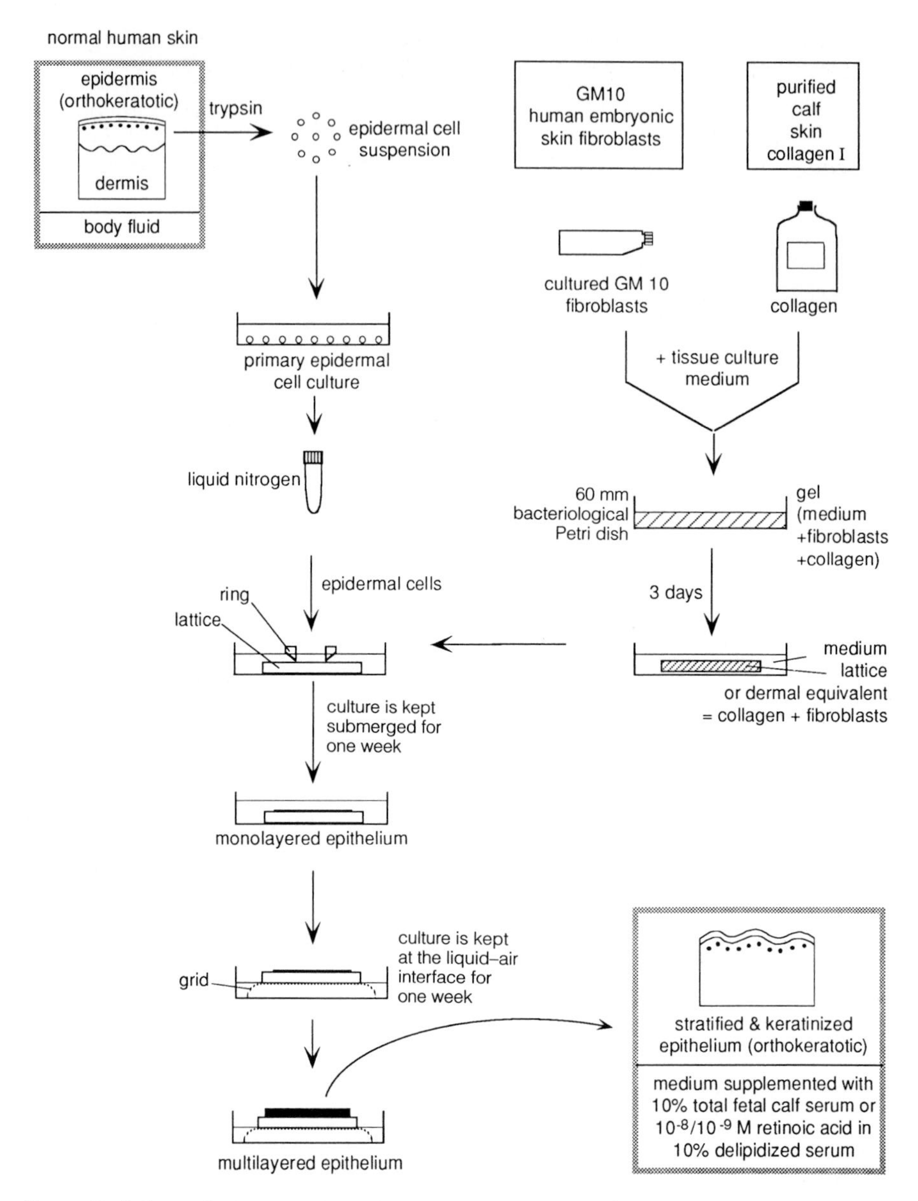

Figure 2. Schematic presentation of the experimental procedure for the reconstruction of human epidermis by culturing keratinocytes on dermal equivalents. From ref. 18 with permission granted by S. Karger A.G., Basel.

models utilizing DED, there is a need for a regular supply of skin which can be difficult to find. Similarly for the source of keratinocytes, the use of surgical samples requires the cooperation of the surgeon, perhaps the consent of the patient and, in any case, a good organization between the supplier and the experimenter. This is why hair follicles are such a convenient source of keratinocytes.

Recently some suppliers of reconstructed skin have appeared on the market (Organogenesis, ATS, Imedex, Mat Tek), but the results are not always reproducible and the prices are often prohibitive.

Protocol 5. Reconstruction of human epidermis by culturing keratinocytes on DED according to Pruniéras *et al.* (3)

Equipment and reagents

- DED (see *Protocol 1*)
- Suspension of human keratinocytes (see *Protocol 4*)
- Culture medium: see *Protocol 2*
- Stainless steel grids (35 mm diameter)[a]
- Stainless steel rings (14 mm diameter)[a]
- Plastic tissue culture dishes (60 mm diameter)

Method

1. Lay the pieces of DED, epidermal side up, on to the bottom of the plastic dishes. Pre-incubate with culture medium for 24 h at 37 °C.
2. Put stainless steel rings on the DED to delineate an area of 1.5 cm^2.
3. Seed 5×10^5 cells (in 0.5 ml culture medium) into the rings.
4. Add culture medium to the dishes so that it covers the DED but *not* the ring. Incubate at 37 °C in an atmosphere of 5% CO_2.
5. After 24 h, remove the rings and place the cultures on to stainless steel grids. Add enough culture medium to reach the underside of the DED. The keratinocytes should remain exposed to the air.
6. Incubate the cultures at 37 °C in a humidified atmosphere containing 5% CO_2. Change the medium twice a week. Usually one week is necessary to obtain both stratification and differentiation of the cultures.

[a] Sterilized by autoclaving.

Protocol 6. Reconstruction of human epidermis by culturing keratinocytes on dermal equivalents according to Bell *et al.* (4) and Asselineau and Pruniéras (10)

Equipment and reagents

- Dermal equivalents (see *Protocol 3*); these should have contracted to a diameter of 18 mm. This requires 2–3 days incubation at 37 °C and an atmosphere containing 5% CO_2
- Suspension of human keratinocytes (see *Protocol 4*)
- Stainless steel rings (14 mm diameter)

Protocol 6. *Continued*

- Lattice culture medium: MEM plus 10% FCS, 0.1% (v/v) penicillin (10 000 IU/ml)–streptomycin (10 000 μg/ml), 1% (v/v) 100 mM sodium pyruvate, 1% (v/v) NEAA, 1% (v/v) 200 mM L-glutamine, 0.4 μg/ml hydrocortisone (400 μl HC stock/500 ml medium), 0.01 μg/ml cholera toxin (5 μl CT stock/500 ml medium), and 10 ng/ml EGF (500 μg EGF stock/500 ml medium) (pH 7.3). For stock solutions see *Protocol 4*
- Stainless steel grids
- Sticking solution: prepare enough solution for sticking all the dermal equivalents. Mix 460 μl MEM (1.76 × concentrated; see *Protocol 3*), 90 μl FCS, 50 μl 0.1 M NaOH, and 100 μl fibroblast culture medium (see *Protocol 3*) for each lattice to be stuck (pH 7.3)
- 3 mg/ml collagen solution (see *Protocol 3*)
- Collagen-coated and uncoated culture dishes
- Scalpel
- Sterile tubes

A. *Sticking the dermal equivalents to the dishes*

1. Aliquot 0.7 ml sticking solution into each sterile tube.
2. Slowly, add 0.3 ml collagen solution down the inner surface of the tubes. Vortex the tubes.
3. Pour the mixture into a culture dish and deposit the lattice on the collagen 'glue' drop, spreading it on the bottom of the dish.
4. Transfer the cultures to an incubator and leave for 30 min.

B. *Seeding and culturing keratinocytes on dermal equivalents*

1. Deposit one stainless steel ring on each lattice to delineate a surface area of 1.5 cm^2.
2. Add 7 ml medium outside each ring. Add 1×10^5 keratinocytes (passage 1) in 0.5 ml medium inside each ring.
3. Incubate the cultures for 2 h in order to allow the cells to attach to the lattice.
4. Change the medium carefully inside the ring and then remove the ring.
5. Seed a suspension of keratinocytes (as described above) into a collagen-coated dish. This culture acts as a control to ensure that the cells reach confluency. The cells on the dermal equivalents can not be viewed using a phase-contrast microscope.
6. Change the medium immersing the lattices twice a week.
7. Once the keratinocytes form a confluent monolayer (usually after one week immersed in culture medium), detach the lattices from the culture dish by cutting the remaining collagen 'glue' with a scalpel.
8. Deposit each lattice on a separate stainless steel grid in a separate tissue culture dish. Add enough medium to reach the bottom of the lattice but leaving the keratinocytes exposed to the air. After one week of air–liquid interface culture, both stratification and differentiation of the cultures should have occurred.

Protocol 7. Reconstruction of human epidermis by culturing hair follicle cells on lattices, according to Lenoir *et al.* (24)

Equipment and reagents

- Human volunteers (with hair!)
- 10 ml collagen–fibroblast gels (see *Protocol 3*, step B)
- Lattice culture medium (see *Protocol 6*)
- Culture medium (see *Protocol 3*)
- Small culture dishes (35 mm diameter)
- Large culture dishes: either 60 mm round dishes or 120 × 120 mm square dishes
- Forceps and scissors

A. *Preparation of the hair follicle explants*

1. Add 1.5 ml culture medium to each of two 35 mm diameter culture dishes.
2. Pluck hair follicles from several areas of the scalp of normal volunteers. Use only follicles in the anagen phase, i.e. those having a visible bulb and sheath.
3. Cut the hair shafts 2 mm above the sheath and immerse the hair follicles in the culture medium contained in the first dish.
4. Using scissors, remove the bulbs from the hair follicles.
5. Cut each hair follicle into two pieces and immerse them in the culture medium contained in the second dish. Five half-follicles are required for each 60 mm diameter lattice.

B. *Preparation of lattices (see Protocol 3)*

1. Prepare 10 ml volume lattices containing 4.6 ml MEM (1.76 × concentrated), 0.9 ml FCS, 0.5 ml 0.1 M NaOH, 0.8 ml 0.1% (v/v) acetic acid, 1 ml fibroblast suspension (containing 4×10^5 cells), and 2.2 ml collagen solution.
2. Let the mixtures gel for 10–15 min in an incubator.

C. *Epidermalization*

1. Plant the explants in an upright position in the freshly cast collagen gels. Plant five half-follicles in 60 mm round dishes or 25 in 120 × 120 mm square dishes. Incubate at 37 °C with 5% CO_2.
2. On day 6, after contraction of the lattices, raise the cultures on to stainless steel grids. Add lattice culture medium to the dishes so that the dermal equivalents are in contact with the medium but the follicles are exposed to humidified air. Reincubate the cultures and change the medium twice a week.

Protocol 7. *Continued*

3. On day 12, remove the hair follicles from the lattices taking care not to damage the growing epithelium. By day 15 the lattices should be completely covered by keratinocytes. Complete differentiation is normally obtained after four weeks (see *Figure 3*).

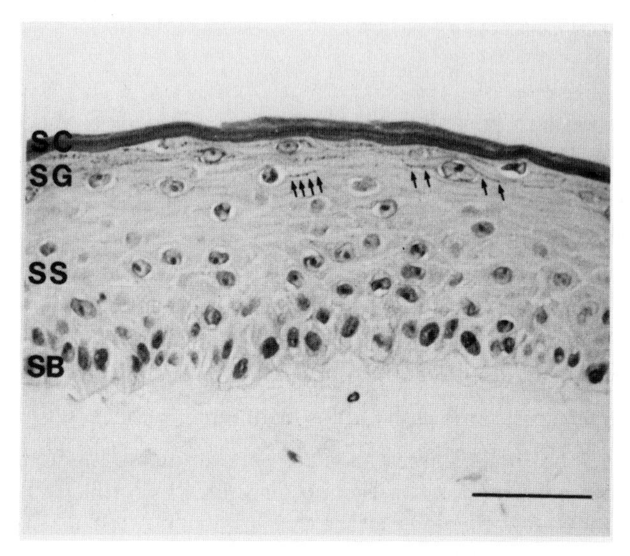

Figure 3. Morphology of reconstructed epidermis obtained after 28 days of culture. (SB, stratum basale; SS, stratum spinosum; SG, stratum granulosum; SC, stratum corneum.) Bar represents 50 μm. From ref. 24 with permission granted by Academic Press.

2.4 Culture media

Since the first relatively simple keratinocyte medium was developed by Rheinwald and Green in 1975, numerous improvements have been made. Simon and Green (15) devised a more complex medium containing a mixture of DMEM and Ham's F-12 media and numerous additives (see *Protocol 4*). However, this medium is not completely chemically defined since it contains fetal calf serum. Successful cultures of human epidermal keratinocytes in a chemically defined medium were first reported in the early 1980s (26–28). Later, studies of the growth and differentiation of keratinocytes in the defined medium (MCDB 153) were published (29–31). Rosdy and Clauss (8) have been able to obtain the terminal differentiation of human keratinocytes grown in chemically defined medium on inert filter substrates at the air–liquid interface.

Protocol 8. Reconstruction of human epidermis by culturing hair follicle cells on DED, according to Lenoir-Viale *et al.* (25)

Equipment and reagents

- Human volunteers
- DED (see *Protocol 1*)
- Stainless steel grids (14 mm diameter)
- Culture medium: as for lattice culture medium (see *Protocol 6*)
- Plastic culture dishes (60 mm diameter)

Method

1. Lay the pieces of DED, epidermal side up, on to the bottom of the plastic dishes. Pre-incubate with culture medium for 24 h at 37 °C.
2. Pluck hair follicles from several areas of the scalp of normal volunteers. Use only follicles in the anagen phase, i.e. having a visible bulb and sheath.
3. Deposit the hair follicles on the DED (three follicles/2 cm^2).
4. Place the pieces of DED (without medium) into an incubator to allow the follicles to attach to the substrate.
5. After 90 min incubation, add one drop of culture medium to each follicle. Reincubate.
6. After two days of culture, add 5 ml of medium to each dish.
7. After six days of culture, elevate the DED to the air–liquid interface using stainless steel grids. Reincubate and change the culture medium twice a week.
8. After thirteen days of culture, remove the hair follicles from the dermis. A multilayered, differentiated epidermis is obtained after a further two days of culture (a total of 15 days of culture).

3. Methods for studying normal cell differentiation

By culturing keratinocytes as described above (see *Protocols 5–8*) under conditions which mimic their natural environment, it is possible to recreate tissues *in vitro* which closely resemble human skin epidermis. These organotypic cell culture models are characterized against their *in vivo* counterparts using the various methods described below (see Sections 3.1–3.4).

3.1 Histology

Standard fixing, sectioning, and staining techniques can be used to allow visualization of the overall organization of the reconstructed tissue including the stratification of the various layers. The basal cells should appear cuboidal with their main axis perpendicular to the dermal–epidermal junction, as in the case of reconstructed epidermis shown in *Figure 3* obtained from hair follicles implanted into collagen–fibroblast lattices. It should also be possible to distinguish the spinous layer, granular layer (with keratohyalin granules), and the cornified layer on the top.

3.2 Electron microscopy

More detailed visualization using electron microscopy is necessary to observe the ultrastructure of the cultures, and in particular to visualize the detailed structure of the basal membrane zone with its different elements (see *Figure 4*). It is also possible to recognize the morphological criteria of terminal epidermal differentiation such as keratohyalin granules, membrane-coating granules, and the keratin pattern of the uppermost corneocytes.

3.3 Immunofluorescence

Cryostat sections can be stained with specific antibodies in order to see the phenotype of the reconstructed epidermis in more detail. Antibodies reacting with basement membrane components can reveal, as in normal human skin, a linear deposition of fibronectin, collagen type IV and laminin at the dermal–epidermal junction, and the polarized localization of the bullous pemphigoid antigen.

Markers specific for keratinocyte terminal differentiation such as the 67 kDa keratin, involucrin, the membrane-bound transglutaminase, and filaggrin are normally expressed in reconstructed epidermis and must have a location similar to that observed in normal human epidermis (see *Figure 5*).

3.4 Protein extraction and identification

Proteins extracted from the cultures can be analysed by one- and two-dimensional gel electrophoresis and be further characterized by immunoblotting. Using these methods, the keratin profile and the processing of profilaggrin into filaggrin can be studied.

4. Uses of *in vitro* reconstructed human skin epidermis

4.1 Studying normal epidermal cell differentiation

With the models described above, the influence of a wide variety of factors including collagens, laminin, culture media, sera, growth factors, air, relative humidity, and chemicals on the differentiation of keratinocytes and epidermal morphogenesis can be tested. The epithelial–mesenchymal interactions can

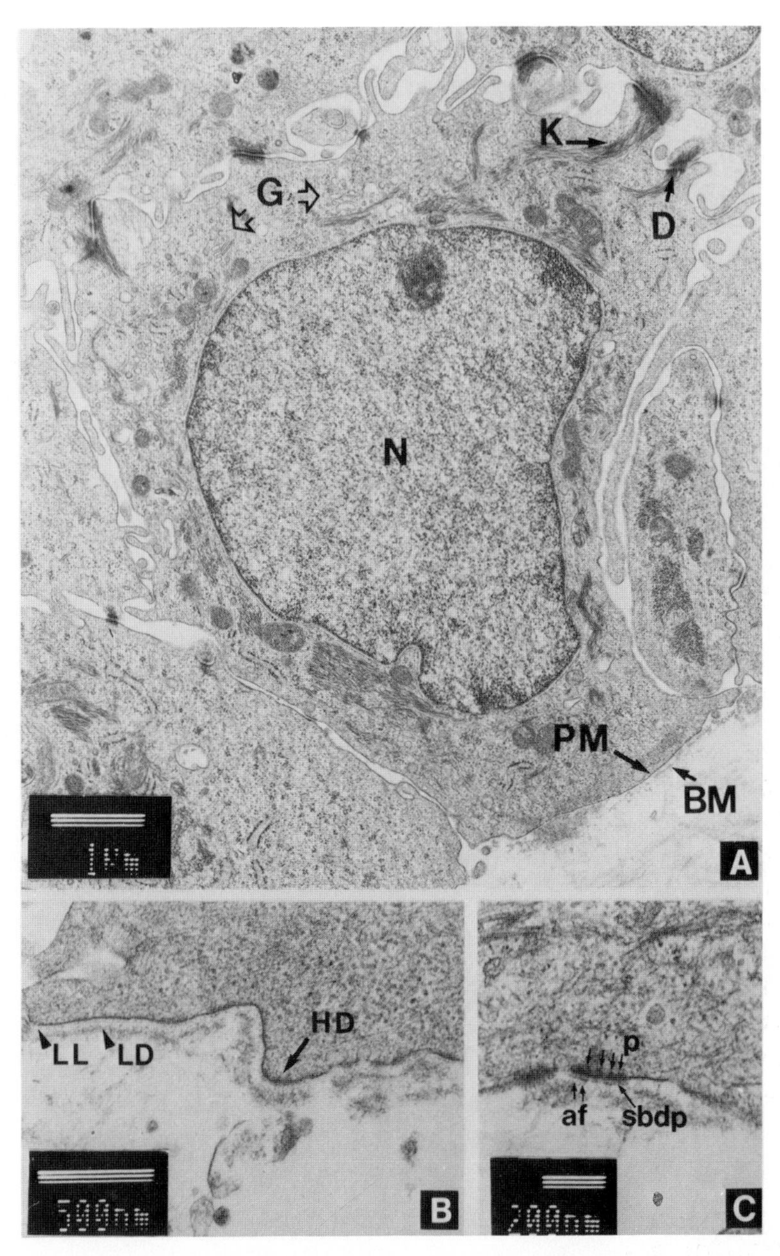

Figure 4. Electron microscopy of sections through a stratified cultured epidermal sheet. (A) A basal cell with basal membrane (BM), plasma membrane (PM), keratin filaments (K), desmosomes (D), and the Golgi apparatus (G) on top of the nucleus (N). Bar represents 1 µm. (B, C) Higher magnification of the basal membrane zone with the lamina lucida (LL), the lamina densa (LD), and the hemidesmosomes (HD). Note the presence of anchoring filaments (af), sub-basal dense plaques (sbdp), and pyramids (p) in the hemidesmosomes. In (B), bar represents 500 nm; in (C), bar represents 200 nm. From ref. 24 with permission granted by Academic Press.

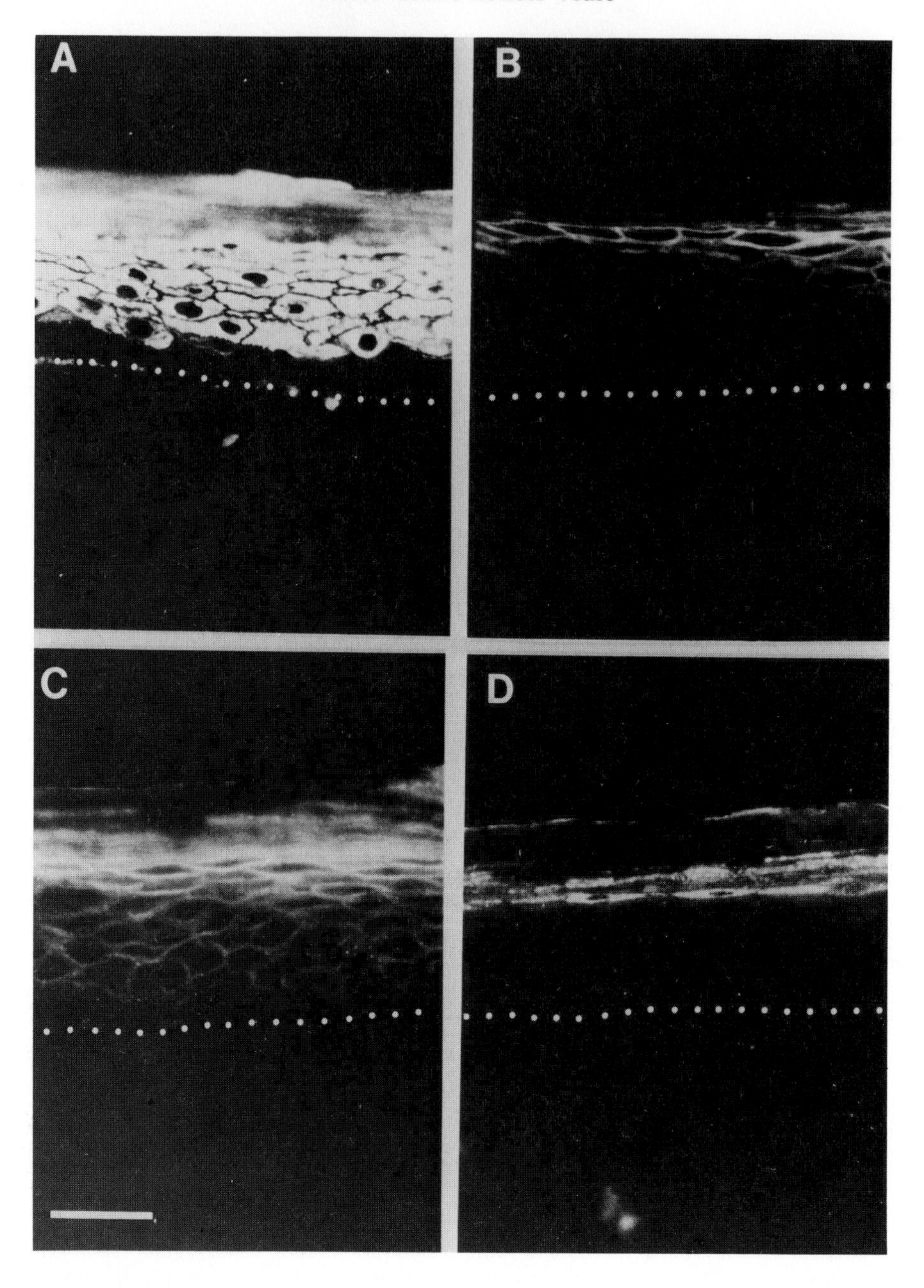

Figure 5. Distribution of the 67 kDa keratin (A), membrane transglutaminase (B), involucrin (C), and filaggrin (D) in a reconstructed epidermis. The dotted line represents the dermal–epidermal junction. Bar represents 25 µm. From ref. 24 with permission granted by Academic Press.

also be studied, and some reports have been published on the role of dermal fibroblasts on epidermal differentiation (32–35). Kinetics of reconstruction can be performed, particularly of the basal membrane zone.

4.2 Studying pathological conditions

Some aspects of pathological conditions have been reproduced by reconstructing human skin *in vitro*. Dermal diseases have been studied, for example epidermolysis bullosa dystrophica (36) and scleroderma (37). By culturing tumoural cells on DED, it has been possible to obtain a tumour-like architecture evoking a Bowen type (38). Seeding Hailey–Hailey keratinocytes on DED gives rise, after 3 weeks of culture, to a tissue resembling patients' epidermis (39). Interactions between psoriatic fibroblasts and keratinocytes have been studied with the dermal equivalent model (40).

4.3 Pharmacological, cosmetic, and pharmaceutical research

The role of retinoic acid in epidermal differentiation and morphogenesis has been extensively studied in reconstructed epidermis *in vitro* (25, 41–44). These studies have significantly contributed to our present understanding of the mechanisms of action of this important mediator (see *Figure 6*). Similarly, but to a lesser extent, the cutaneous effects of vitamin D have also been studied using organotypic models (45).

One major advantage of these three-dimensional cultures is that they permit the topical application of drugs (see *Figure 7*). The presence of the cornified layer, although imperfect, allows evaluation of the epidermal-barrier function, and comparisons have been made with normal human skin (46, 47).

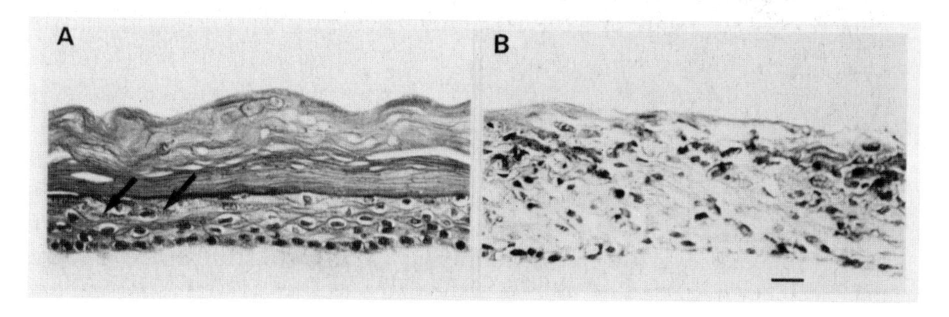

Figure 6. Morphology of the epithelium formed by human keratinocytes grown for two weeks on a dermal equivalent and either emerged for one week in the absence of retinoic acid (A) or in the presence of 1 μM retinoic acid (B). Note in A the presence of keratohyalin granules above the basal layer (arrows) and particularly numerous horny layers. Bar, 30 μm. From ref. 41 with permission granted by Academic Press.

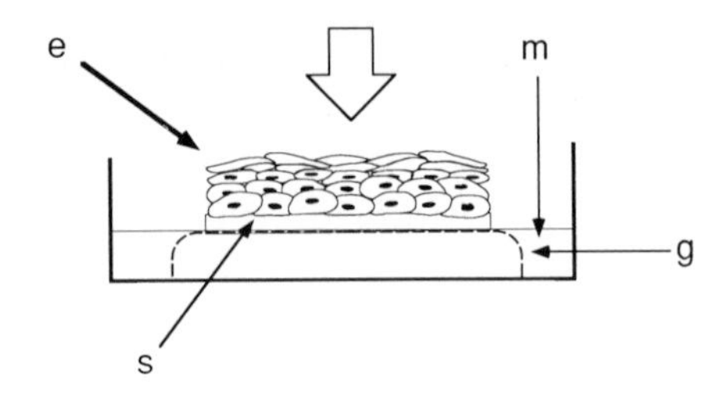

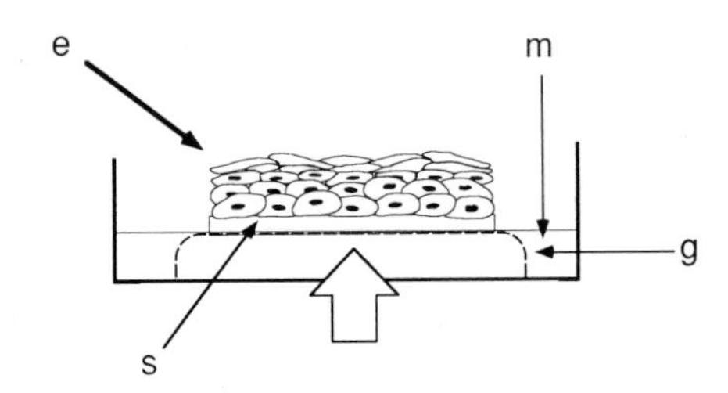

Figure 7. Epidermis (e) reconstructed on dermal substrate (s) exposed to the atmosphere to study the effects of pharmacological agents administered either 'systematically' or 'topically' (arrows). m, culture medium; g, grid. From ref. 18 with permission granted by S. Karger, A.G. Basel.

Skin equivalents are more and more frequently being used in the cosmetic industry to replace laboratory animals. For example, the potential irritancy of surfactants has been evaluated using three epithelial parameters: viability, barrier function, and the release of an inflammatory mediator, interleukin 1α (48). A good correlation was found with the data obtained *in vivo*.

5. Conclusions

Reconstruction of human epidermis *in vitro* may help basic researchers to better understand the biology of keratinocytes and the mechanisms involved in normal and pathological epidermal morphogenesis. Industrial scale-production of these models is making them more and more easily available and reproducible. Thus, a promising tool for screening purposes as well as for improving the knowledge of dermal pharmacotoxicology is now within reach. Grafting of these *in vitro* reconstructed skin models has been tried on humans, but for the moment, only preliminary results have been obtained and the technique of Rheinwald and Green (1, 2) for obtaining intact epidermal sheets is still the method of choice.

Acknowledgements

I wish to thank my colleagues D. Asselineau and M. Régnier who passed on their experience to me and gave me all their protocols; B. A. Bernard with whom I performed the work on reconstructed epidermis from hair follicles. I thank also S. Michel and W. R. Pilgrim for critical reading of this manuscript. I am grateful to M. Simicic for preparing the manuscript and to D. Poisson for artwork.

References

1. Rheinwald, J. G. and Green, H. (1975). *Cell*, **6**, 317.
2. Rheinwald, J. G. and Green, H. (1975). *Cell*, **6**, 331.
3. Pruniéras, M., Régnier, M., and Schlotterer, M. (1979). *Ann. Chir. Plast.*, **24** (4), 357.
4. Bell, E., Ivarsson, B., and Merrill, C. (1979). *Proc. Natl Acad. Sci. USA*, **76** (3), 1274.
5. Boyce, S. T. and Hansbrough, J. F. (1988). *Surgery*, **103**, 421.
6. Tiollier, J., Dumas, H., Tardy, M., and Tayot, J. L. (1990). *Exp. Cell Res.*, **191**, 95.
7. Shahabeddin, L., Berthod, F., Damour O., and Collombel, C. (1990). *Skin Pharmacol.*, **3** (2), 107.
8. Rosdy, M. and Clauss, L. C. (1990). *J. Invest. Dermatol.*, **95**, 409.
9. Freeman A. E., Igel, H. J., Herrman, B. J., and Kleinfeld, F. L. (1976). *In Vitro*, **12**, 352.
10. Asselineau, D. and Pruniéras, M. (1984). *Br. J. Dermatol.*, **3** (Suppl. 27), 219.
11. Rénier, M., Pruniéras, M., and Woodley, D. (1981). In *Frontiers of matrix biology* (ed. L. Robert, Créteil), Vol. **9**, pp. 4–35. Karger, Basel.
12. Yannas, I. V., Burke, J. F., Gordon, P. L., Huang, C., and Rubenstein, R. H. (1980). *J. Biomed. Mater. Res.*, **14**, 65.
13. Yannas, I. V., Burke, J. F., Gordon, P. L., Huang, C., and Rubenstein, R. H. (1980). *J. Biomed. Mater. Res.*, **14**, 107.
14. Rheinwald, J. G. (1989). In *Cell growth and division: a practical approach* (ed. Renato Baserga), pp. 81–94. IRL Press, Oxford.
15. Simon, M. and Green, H. (1985). *Cell*, **40**, 677.
16. Vermorken, A. J. M. (1985). *Altern. Lab. Anim.*, **13**, 8.
17. Limat, A. and Noser, F. K. (1986). *J. Invest. Dermatol.*, **87**, 485.
18. Régnier, M., Asselineau, D., and Lenoir, M. C. (1990). *Skin Pharmacol.*, **3**, 70.
19. Coulomb, B., Saiag, P., Bell, E., Breitburd, F., Lebreton, C., Heslan, M., and Dubertret, L. (1986). *Br. J. Dermatol.*, **114**, 91.
20. Basset-Séguin, N., Culard, J. F., Kerai, C., Bernard, F., Watrin, A., Demaille, J., and Guilhou, J. J. (1990). *Differentiation*, **44**, 232.
21. Noser, F. K. and Limat, A. (1987). *In Vitro Cell. Dev. Biol.*, **23** (8), 541.
22. Lenoir, M. C. and Bernard, B. A. (1990). *Skin Pharmacol.*, **3** (2), 97.
23. Bailly, C., Drèze, M. S., Asselineau, D., Nusgens, B., Lapière, C. M., and Darmon, M. (1990). *J. Invest. Dermatol.*, **94**, 47.
24. Lenoir, M. C., Bernard, B. A., Pautrat, G., Darmon, M., and Shroot, B. (1988). *Dev. Biol.*, **130**, 610.

25. Lenoir-Viale, M. C., Galup, C., Darmon, M., and Bernard, B. A. (1993). *Arch. Dermatol. Res.*, **285**, 197.
26. Peehl, D. M. and Ham, R. G. (1980). *In Vitro*, **16**, 526.
27. Tsao, M. C., Walthall, B. J., and Ham, R. G. (1982). *J. Cell Physiol.*, **110**, 219.
28. Boyce, S. T. and Ham, R. G. (1983). *J. Invest. Dermatol.*, **81**, 33S.
29. Boyce, S. T. and Ham, R. G. (1986). In *In vitro models for cancer research* (ed. M. M. Weber and L. I. Sekely), Vol. 3, pp. 245–74. CRC Press, Boca Raton, FL.
30. Wille, J. J., Pittelkow, M. R., Shipley, G. D., and Scott, R. E. (1984). *J. Invest. Dermatol.*, **121**, 44.
31. Pittelkow, M. R., Wille, J. J., and Scott, R. E. (1986). *J. Invest. Dermatol.*, **86**, 410.
32. Coulomb, B., Lebreton, C., and Dubertret, L. (1989). *J. Invest. Dermatol.*, **92**, 122.
33. Krejci, N. C., Cuono, C. B., Langdon, R. C., and McGuire, J. (1991). *J. Invest. Dermatol.*, **97**, 843.
34. Saintigny, G., Bonnard, M., Damour, O., and Collombel, C. (1993). *Acta Dermatol. Venereol.*, **73**, 175.
35. Smola, H., Thiekötter, G., and Fusenig, N. E. (1993). *J. Cell Biol.*, **122** (2), 417.
36. Ehrlich, H. P., Buttle, D. J., Trelstad, R. L., and Hayashi, K. (1983). *J. Invest. Dermatol.*, **80**, 56.
37. Maquart, F. X., Gillery, P., Kalis, B., and Borel, J. P. (1986). *J. Invest. Dermatol.*, **87**, 154.
38. Régnier, M., Desbas, C., Bailly, C., and Darmon, M. (1988). *In Vitro Cell. Dev. Biol.*, **24**, 625.
39. Régnier, M., Ortonne, J. P., and Darmon, M. (1990). *Arch. Dermatol. Res.*, **281**, 538.
40. Saiag, P., Coulomb, B., Lebreton, C., Bell, E., and Dubertret, L. (1985). *Science*, **230**, 669.
41. Asselineau, D., Bernard, B. A., Bailly, C., and Darmon, M. (1989). *Dev. Biol.*, **133**, 322.
42. Asselineau, D., Dale, B. A., and Bernard, B. A. (1990). *Differentiation*, **45**, 221.
43. Kopan, R., Traska, G., and Fuchs, E. (1987). *Cell Biol.*, **105**, 427.
44. Régnier, M. and Darmon, M. (1989). In *Vitro Cell. Dev. Biol.*, **25** (11), 1000.
45. Régnier, M. and Darmon, M. (1991). *Differentiation*, **47**, 173.
46. Ponec, M., Wauben-Penris, P. J. J., Burger, A., Kempenaar, J., and Boddé, H. E. (1990). *Skin Pharmacol.*, **3** (2), 126.
47. Régnier, M., Caron, D., Reichert, U., and Schaefer, H. (1992). *Skin Pharmacol.*, **5**, 49.
48. Roguet, R., Cohen, C., Rougier, A., and Dossou, K. G. (1993). *Nouv. Dermatol.*, **12**, 468.

A1

Addresses of suppliers

Advanced Tissue Sciences (ATS), 10933 North Torrey Pines Road, La Jolla, CA 92037, USA.

Agar Scientific Ltd, 66a Cambridge Road, Stansted, Essex, CM24 8DA, UK.

Aldrich Chemical Co., The Old Brickyard, New Road, Gillingham, Dorset, SP8 4BR, UK.

American Type Culture Collection (ATCC), 12301 Parklawn Drive, Rockville, MD 20852, USA.

Amersham International, Little Chalfont, Bucks, HP7 9NA, UK; Amersham France, 12 avenue des Tropiques, BP 144, 91944 Les Ulis Cedex, France.

Antibodies, Inc., PO Box 1560, Davis, CA 95617, USA.

ATGC, ZI des Richardets, 24 Rue du Ballon, 93160 Noisy le Grand, France.

ATS (Advanced Tissue Sciences), 10933 North Torrey Pines Road, La Jolla, CA 92037, USA.

Baxter Diagnostics Inc., Scientific Products Division, 1430 Waukegan Road, McGraw Park, IL 60085, USA.

Bayer plc, Evans House, Hamilton Close, Basingstoke, Hants, RG21 2YE, UK.

BDH: see Merck.

Bearing Inc., 4301 Dacoma Street, Houston, TX 77092, USA.

Beckman Instruments, 92–94 chemin des Bourdons, 93220 Cagny, France.

Becton–Dickinson UK Ltd, Betweens Towns Road, Cowley, Oxford, OX4 3LY, UK; Becton Dickinson, 2350 Qume Drive, San Jose, CA 95131, USA; Becton–Dickinson (Falcon) 2 Bridgewater Lane, Lincoln Park, NJ 07035, USA; Becton–Dickinson Labware, 1950 William Drive, Oxnard, CA 93030, USA; Falcon/Becton Dickinson, 1 Becton Drive, Franklin Lakes, NJ 07417, USA.

The Binding Site, 5889 Oberlin Drive, San Diego, CA 92121, USA.

Biocell Research Laboratories, Cardiff Business Technology Centre, Senghenydd Road, Cardiff, CF2 4AY, Wales.

Biogenesis, 104 Little Mill Road, Sandown, NH 03873, USA.

Biogenex Laboratories, 4600 Norris Canyon Road, San Ramon, CA 94583, USA.

Biorad Laboratories Ltd, Biorad House, Marylands Avenue, Hemel Hempstead, Herts HP2 7TD, UK; Bio-Rad Laboratories, Life Sciences Group, 2000 Alfred Nobel Drive, Hercules, CA 94547, USA.

Boehringer–Mannheim GmbH Biochemica, Postfach 31 01 20, D-6800 Mannheim 31, Germany; Boehringer–Mannheim, Biochemicals Corp, Inc, PO Box 50816, Indianapolis, IN 46250, USA; Boehringer–Mannheim, UK (Diagnostics & Biochemicals) Ltd, Bell Lane, Lewes, East Sussex BN7 1LG UK.

Calbiochem–Novabiochem (UK) Ltd, Unit C2A, Boulevard Industrial Park, Beeston, Notts, NG9 2JR, UK.

Cell Line Associates Inc., PO Box 35, Newfield, NJ 08344, USA.

Central Laboratory of the Blood Transfusion Service, Rode Kruis, Plesmanlaan 125, 1066 CX Amsterdam, The Netherlands.

Chemicon International Inc., 28835 Single Oak Drive, Temecula, CA 92590, USA.

Clonetics Corporation, 9620 Chesapeake Drive, San Diego, CA 92123, USA.

Collaborative Biomedical Products, 2 Oak Park, Bedford, MA 01730, USA; Universal Biological Ltd, 12–14 St Anns Crescent, London, SW18 2LS, UK.

Collaborative Research Inc., Biomedical Products Division, Two Oak Park, Bedford, MA 01730, USA; Collaborative Research Inc, Research Products Division, 128 Spring Street, Lexington, MA 02173, USA; Collaborative Research Inc, 1365 Main Street, Waltham, MA 02154, USA.

Corex, Bioblock Scientific, BP 111, 67403 Illkirch, France.

Costar UK Ltd, 10 The Valley Centre, Gordon Road, High Wycombe, Bucks, HP13 6EQ, UK; Costar UK, Victoria House, 28–39 Desborough Street, High Wycombe, Bucks, HP11 2NF; Costar Corporation, 1 Alewife Center, Cambridge, MA 02140, USA; Costar Europe Ltd, PO Box 94, 1170 AB Badhoevedorp, The Netherlands.

DAKO Ltd, 16 Manor Courtyard, Hughendon Avenue, High Wycombe, Bucks, HP13 5RE, UK; DAKO A/S, Produktionsvej 42, PO Box 1359, DK-2600 Glostrup, Denmark; DAKO Corporation, 6392 Via Real, Carpinteria, CA 93013, USA.

Detrona, Ekholmsvägen 87, 58262 Linköping, Sweden.

Developmental Studies Hybridoma Bank, University of Iowa, Department of Biological Sciences, Iowa City, IA 52242, USA.

Ethicon GmbH, Robert-Koch-Str 1, D-2000 Norderstedt, Germany.

Europath Ltd, Highland Comfort, Union Hill, Stratton, Bude, EX23 9BL, UK.

European Collection of Animal Cell Cultures (ECACC) Division of Biologies, PHLS Centre for Applied Microbiology and Research, Porton Down, Salisbury SP4 0JG, UK.

E. Y. Laboratories Inc., PO Box 1787, San Mateo, CA 94401, USA.

Falcon, 47 avenue Marie-Reynoard, 38100 Grenoble, France.

Fisher Scientific, 10700 Rockley Road, Houston, TX 77251-1307, USA.

Fluka Chemicals Ltd, The Old Brickyard, New Road, Gillingham, Dorset, SP8 4JL, UK; Fluka S.a.r.l., L'Isle d'Abeau Chesnes, BP 701, F-38297, St Quentin Fallavier, Cedex France.

Forma Scientific, Scientific BV, De Meern, The Netherlands.

Gelman Sciences Ltd, 10 Arrowden Road, Brackmills, Northampton, NN4 0EZ, UK; Gelman Sciences SA, Parc-Club de la Haute Maison, 16 rue Galilée, Citée Descartes, 77420 Champs-sur-Marne, France.

GenProbe, Inc., 9880 Campus Point Drive, San Diego, CA 92121, USA.

Gibco–BRL, Life Technologies Ltd, PO Box 35, Trident House, Renfrew Road, Paisley, PA3 4EF, Scotland; GIBCO/BRL Laboratories, Life Technologies Inc, PO Box 68, 3175 Staley Road, Grand Island, NY 14072, USA; Gibco BRL/Life Technologies, PO Box 6009, Gaithersburg, MD 20884, USA; Gibco Brl, 1 rue du Limousin, 95 310 Saint-Ouen-l'Aumône, France; Life Technologies SARL (GIBCO), ZAC des Bellevies, Avenue du Gros Chêne, 95160 Eragny, France.

Greiner GmbH, Maybachstrasse, PO Box 1162, D-72632 Frickenhausen, Germany; Greiner Labortechnic Ltd., 13 Station Road, Cam, Dursley, Gloucs GL11 5NS, UK.

Hoefer Scientific Instruments, 654 Minnesota Street, San Francisco, CA 94107, USA.

Hyclone Laboratories, Inc., 1725 South Hyclone Road, Logan, Utah 84321-9965, USA.

ICN Biochemicals, PO Box 28050, Cleveland, OH 44128, USA.

ICN Biomedicals, PO Box 5023, 3300 Hyland Avenue, Costa Mesa, CA 92626, USA.

ICN–Flow Laboratories Ltd (Titertec/Flow), PO Box 17, Irvine, Ayrshire KA 12 8NB, Scotland; Flow Laboratories, 99 rue de la République, 92800 Puteaux, France; ICN–Flow Inc, 7655 Old Springhouse Road, McLean, VA 22102, USA.

IEC, 300 Second Avenue, Needham Heights, MA 02194, USA.

Imedex, BP 38, 69630 Chaponost, France.

Imperial Cancer Research Fund, Clare Hall Laboratories, South Mimms, Potters Bar, Herts, UK.

Institut Jacques Boy SA, 45 rue Cognacq Jay, B P 1430, 51065 Reims, Cedex, France.

J. Bio, P-91967 Les Ulis, France.

Kao Corporation, 1-14-10 Nihombashi Kayabacho Chuo-ku, Tokyo, Japan.

Labsystems/Life Sciences International Ltd, Unit 5, The Ringway Centre, Edison Road, Basingstoke, Hants, RG1 2YH, UK.

Leitz, Wild Leitz USA, Inc., 24 Link Drive, Rockleigh, NJ 07647, USA.

Marathon Laboratory Supplies, Unit 6, 55–57 Park Royal Road, London, NW10 7CP, UK.

Leo Pharmaceutical Products B.V., Pampuslaan 186, 1382 JS Weesp, The Netherlands.

Mat Tek Corporation, 200 Homer Avenue, Ashland, MA 01721, USA.

Merck, Frankfuter Str 250, Postfach 4119, D-6100 Darmstadt, Germany.

Merck Ltd (BDH Laboratory Supplies), Merck House, Poole, Dorset, BH15 1TD, UK.

Millipore (UK) Ltd, The Boulevard, Blackmoor Lane, Watford, Herts, WD1 8YW, UK; Millipore (UK) Ltd, Millipore House, 11–15 Peterborough Road, Harrow, Middlesex, HA1 2BR, UK; Millipore Corp, 80 Ashby Road, Bedford, MA 01730 USA; Millipore Corporation, PO Box 1962, 397 William Street, Marlborough, MA 01752-1962, USA; Millipore SA, 39 route Industrielle de la Hardt, 67120 Molsheim, France.

Molecular Probes, PO Box 22010, 4849 Pitchford Avenue, Eugene, OR 97402, USA.

Moria, 108 boulevard Saint-Germain, 75006, Paris, France.

NEN Research Products, 549 Albany Street, Boston, MA 02118, USA.

New England Nuclear/Du Pont, Burky Mill Plaza, P-24, Wilmington, DE 19898, USA.

Nordic Immunological Laboratories b.v., PO Box 22, NL-5000 AA Tilburg, The Netherlands.

Northumbria Biologicals Ltd, South Nelson Industrial Estate, Cranlington, Northumberland, NE23 9HL, UK.

NPBI, Plesmanlaan 125, NL-1066 CX, Amsterdam, The Netherlands.

Nucleopore Corporation, 7035 Commerce Drive, Pleasanton, CA 94566-3294, USA; Nucleopore Co, D-7400 Tubingen, Falkenweg 47, Germany.

Nunc A/S, Postbox 280, Kamstrupvej 90, Kamstrup, DK-4000 Roskilde, Denmark; Nunc Inc, 200 N. Aurora Road, Naperville, IL 60566, USA.

Olympus, OSI, BP 124, 78312 Maurepas cedex, France.

Organogenesis, Inc., 83 Rogers Street, Cambridge, MA 02142, USA.

Pharmacia LKB Biotechnology AB, Björkgatan 30, S-75182 Uppsala, Sweden; Pharmacia LKB Biotechnology, PO Box 1327, Piscataway, NJ 08855, USA.

Polyscience Inc., Park Scientific Ltd, 24 Lowfarm Road, Moulton Park, Northampton, NN3 1HY, UK.

Pierce, Life Science Laboratories Ltd, Sedgewick Road, Luton, LU1X 9DT, UK.

Sanbio b.v., PO Box 540, NL-5400 AM Uden, The Netherlands.

Sandoz, CH-4002, Basel, Switzerland; Sandoz Research Institute, Brunnerstrasse 59, A-1235, Vienna, Austria.

Sera-Lab Ltd, Crawley Down, Sussex, RH10 4FF, UK.

Seromed: *see* ATGC.

Shandon HPLC, Chadwick Road, Astmoor, Runcorn, Cheshire, WA7 1PR, UK.

Sigma Chemical Co. Ltd, Fancy Road, Poole, Dorset, BH17 7NH, UK; SIGMA (Aldrich) Chimie, L'Isle d'Abeau Chesnes, BP 701, 38297 Saint Quentin Fallavier, Cedex, France; Sigma Chemical Co, PO Box 14505 (for orders outside of USA), PO Box 14508 (USA), St Louis, MO 63178, USA.

Sterilin Ltd, 43–45 Broad Street, Teddington, Middlesex, TW11 8QZ, UK.

Taab Laboratories Ltd, Unit 3, Minerva House, Colleva Industrial Park, Aldermaston, Berks, RG7 8NA, UK.
Techne (Cambridge) Ltd, Duxford, Cambridge, CB2 4PZ, UK.
Thamer, Postbus 233, 1420 AE, Uithoorn, The Netherlands.
Transbio, 91 avenue J.B. Clémont, 92100 Boulogne, France.
Vector Laboratories Inc., 30 Ingold Road, Burlingame, CA 94010, USA.
World Precision Instruments (WPI), 375 Quinnipiac Avenue, New Haven, CT 06513, USA.
Worthington Biochemical Corporation, Halls Mill Road, Freehold, NJ 07728, USA; Lorne Biochemicals, Twyford, Reading, Berks, RG10 9NL, UK.
Zymed Laboratories Inc., 458 Carlton Court, So. San Francisco, CA 94080, USA.

A2

An *in vitro* model of the blood–brain barrier for studying drug transport to the brain

A. G. DE BOER, H. E. DE VRIES, P. J. GAILLARD, and D. D. BREIMER

1. Introduction

The blood–brain barrier (BBB) comprises the endothelial lining of the microvessels in the brain and is considered to be the main barrier to drug transport into the CNS; the BBB has a surface area 5000 times larger than that presented by the blood–cerebrospinal fluid barrier (1). The endothelial cells of the BBB possess tight junctions, low pinocytotic activity, no fenestrae, and abundant mitochondria, and represent a formidable metabolic, as well as a physical barrier, to many labile, hydrophilic CNS drugs (see *Figure 1*;ref. 2). It has been shown in co-culture systems that the integrity of the tight

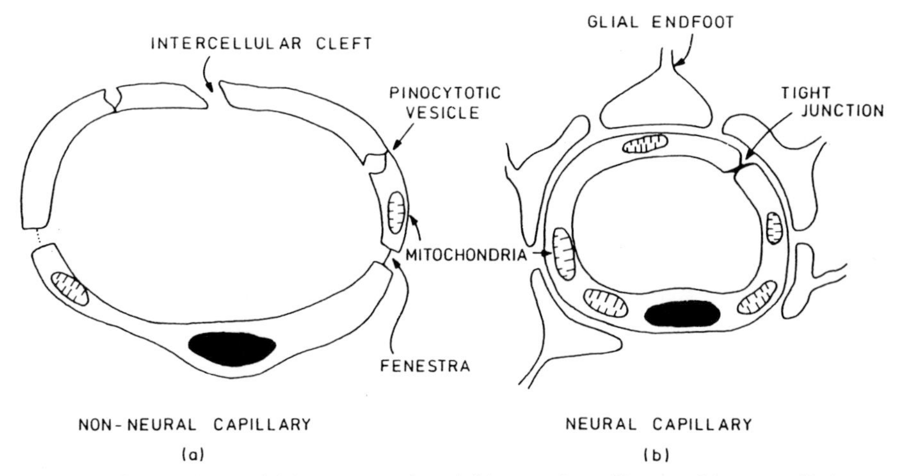

Figure 1. Comparison of (a) non-neural and (b) neural capillary architecture. Redrawn from ref. 2.

junctions of cerebrovascular endothelial cells is influenced by astrocytes (3–5). Pericytes are also attracted by endothelial cells (6), and neurones and some blood-borne factors (e.g. hormones) can affect the functionality of the BBB (7). In addition, the endothelial lining of the circumventricular organs (CVOs) is relatively leaky (8, 9) which indicates that the BBB is not a homogeneous system.

Brain capillary endothelial cells can be isolated by various procedures including completely non-enzymatic (10), combined mechanical–enzymatic (11), or solely enzymatic methods (12, 13). We use the technique described by Audus and Borchardt (ref. 13; see *Protocol 1*) which comprises a two-step enzymatic procedure with subsequent density-gradient centrifugation steps to separate the cells from contaminating neuronal debris and other cells (e.g. pericytes and fibroblasts). The isolated cells can be cultured on solid plastic or, in order to model the BBB, on porous membranes (e.g. Transwell; Costar). The cells can also be cultured with astrocytes or in astrocyte-conditioned medium (ACM). Astrocytes can be isolated from rat pups according to the procedure of Tio *et al.* (ref. 14; see *Protocol 2*).

Protocol 1. Isolation, freezing, and culture of bovine brain microvessel endothelial cells (BMECs)

Equipment and reagents

- Two bovine (calf) brains: transported fresh from the slaughterhouse in ice-cold transport medium[a]
- Transport medium: MEM (D-val modification) containing 50 mM Hepes, 167 IU/ml penicillin, 100 μg/ml streptomycin, 50 μg/ml polymyxin B, and 2.5 μg/ml amphotericin B (pH 7.4)
- Incubation medium I: 500 ml of transport medium containing 2.5 g of dispase.[b] This is enough for two brains
- Incubation medium II: 50 ml transport medium containing 1–4 mg/ml collagenase/dispase (Boehringer–Mannheim; adjusted according to individual batches)
- 13% (w/v) dextran solution (average M_w, 70 000; Sigma) containing 167 IU/ml penicillin and 100 μg/ml streptomycin
- Temperature controlling freezing machine (e.g. CRYOMED, Forma Scientific) and cryonials
- Collagen/fibronectin-coated filters: Transwell filters (Costar) treated with rat tail collagen (0.1% (v/v) in acetic acid or commercial type I collagen) and fibronectin (0.2 mg/ml solution, 10 μg/cm^2)
- Culture medium: 4.85 g/litre MEM powder, 5.35 g/litre Ham's F-12 powder, 10 mM, Hepes, 167 IU/ml penicillin, 100 μg/ml streptomycin, 9% (v/v) plasma-derived horse serum (Hyclone), 50 μg/ml polymyxin B sulfate, and 2.5 μg/ml amphotericin B. Adjust the pH of the medium to pH 7.4 using 10 M NaOH. Add the horse serum immediately before use and, once added, do not store the medium for more than two weeks
- MEM (pH 7.4)
- 4 × 75 ml Percoll gradient
- Freezing medium: a 42:5 mixture of culture medium and DMSO

A. *Isolation of BMEC*

1. Perform the isolation procedure in a laminar down-flow cabinet and ensure that you use sterile equipment and sterile gloves.
2. Remove and discard the meninges. Using a razor blade or curved forceps, collect the grey matter in transport medium.

3. Mince the material for 10–20 s in a blender until a change in viscosity can be observed. Make the volume up to 500 ml using incubation medium I.
4. Incubate the suspension for 30 min at 37 °C. Adjust the pH with MEM (pH 9) to pH 7.4.
5. Further incubate the mixture with gentle shaking (200 r.p.m.) in a waterbath for 2 h at 37 °C.
6. Centrifuge the mixture at 1000 *g* for 12 min at 4–10 °C.
7. Resuspend the pellet in 500 ml of 13% dextran solution and centrifuge at 5800 *g* for 15 min at 4–10 °C.
8. Resuspend the pellet in 20 ml of incubation medium II and incubate the suspension in a shaking waterbath (200 r.p.m.) for 4.5 h at 37 °C.
9. Dilute the capillary suspension up to 50 ml with MEM and centrifuge at 1000 *g* for 10 min at room temperature.
10. Resuspend the pellet in 8 ml of MEM.
11. Place 2 ml aliquots of the capillary suspension each on to 50 ml of a 50% continuous Percoll gradient and centrifuge at 1000 *g* for 10 min at room temperature.
12. Resuspend the pellet in culture medium and centrifuge at 600 *g* for 10 min at room temperature.

B. *Freezing and thawing BMEC*

1. Resuspend the pellet in freezing medium and centrifuge at 600 *g* for 10 min at room temperature.
2. Resuspend the pellet in 10 ml of freezing medium and prepare 2 ml aliquots in cryovials.
3. Freeze the cell suspensions at a rate of 1 °C/min to –80 °C using a suitable freezing apparatus.
4. Transfer the vials to liquid nitrogen.
5. When you want to use the cells, rapidly thaw the appropriate number of aliquots.
6. Add culture medium to the cells and centrifuge at 600 *g* for 10 min at room temperature. Repeat this step once.

C. *Culture of BMEC*

1. Seed 5×10^4 cells/cm^2 into either tissue culture plates or collagen/fibronectin-coated filters.
2. Incubate the cells at 37 °C in an atmosphere of 5% CO_2/95% air and 100% humidity.[c]

Protocol 1. *Continued*

3. Renew the culture medium every three days until the cells are confluent.[d, e]

[a] The animals are dead for about 10 min before removal of the brains. In our laboratory, the isolation of BMEC normally occurs within 1 h of the death of the animals.
[b] The dispase should be dissolved in a small volume of medium and paper filtered into the remaining medium. The activity of dispase varies considerably between batches and so the amount given is only approximate.
[c] Contamination of the cell culture by unwanted cell types (primarily pericytes) can be considerably reduced by refeeding the culture 30 min after seeding.
[d] Angiogenesis can be reduced by seeding at a high (confluent) cell density and by avoiding the addition of growth factors.
[e] The TER across a confluent monolayer of BMEC can be increased by culturing the cells in astrocyte-conditioned medium and by applying the methods published by Rubin *et al.* (15).

Protocol 2. Isolation and culture of astrocytes[a]

Equipment and reagents

- Hanks' buffer: 0.14 M NaCl, 5 mM KCl, 0.3 mM Na_2HPO_4, 0.4 mM KH_2PO_4, and 25 mM Hepes. Adjust to pH 7.4
- 0.05% (w/v) trypsin/0.02% (w/v) EDTA (Gibco–BRL)
- Nylon sieve mesh (120 μm and 45 μm)
- Culture medium: MEM supplemented with 0.2% (w/v) glutamine, 10% (v/v) fetal calf serum, 50 mM Hepes, 167 IU/ml penicillin, and 100 μg/ml streptomycin (pH 7.4)
- Sterile dissecting instruments

Method

1. Dissect the cortices and remove the meninges.
2. Mince the meninges and incubate them in Hanks' buffer at 37 °C in a shaking waterbath (80 r.p.m.) for 15 min.
3. Add the trypsin/EDTA solution until the final concentration is 0.1% (v/v) and incubate the suspension for 25 min at 37 °C in a shaking waterbath (80 r.p.m.).
4. Filter the mixture through a 120 μm nylon sieve mesh. Wash the filtrate and filter it through a 45 μm nylon sieve mesh.
5. Culture the isolated astrocytes in culture medium.
6. Remove the oligodendrocytes after 8 days by shaking the cultures in a 37 °C waterbath overnight at 120 r.p.m. in MEM (16).[b]
7. Collect the ACM from the cultures after 48 h of conditioning.[c]

[a] Astrocytes are isolated from rat pups of up to 2-days-old.
[b] Astrocytes can be characterized by expression of glial fibrillary acidic protein. Purity should be ≥ 95%.
[c] Store the ACM at –20 °C.

2. Characterization of BMEC

The expression of various features can be used to identify positively isolated microvessels and BMEC in culture (17):

(a) Electron microscopy can be used to identify features inherent to these cells, e.g. tight junctions.

(b) A high transendothelial electrical resistance (TER) across confluent monolayers of BMEC indicates the presence and integrity of tight junctions. It can be determined conveniently by using commercially available apparatus (see Chapter 6, section 2.3.1) or by using the four electrode system described by Hurni *et al.* (18).

(c) Enzymes which are expressed by microvessel endothelial cells but not by their neighbouring cells, e.g. alkaline phosphatase and γ-glutamyl transpeptidase (γ-GTP), can be used to identify BMEC. However, it has been demonstrated by Risau *et al.* (19) that pericytes also express γ-GTP. *Table 1* demonstrates that the amount of γ-GTP expressed by BMEC varies according to the culture time, the use of ACM, and the presence or absence of astrocytes.

(d) General markers of endothelial cells, e.g. angiotensin-converting enzyme (ACE), Factor VIII/von Willebrand antigen, and DiI-acetyl-LDL (20) can be used to identify BMEC (see Chapter 5). Researchers should be aware, however, that some of these markers may be progressively lost during culture. It is generally advisable to use primary cell cultures.

3. Applications of BMEC

BMEC monolayers can be used to study the paracellular and transcellular transport of drugs across the BBB (21); the TER can be measured as an indication of changes in paracellular permeability. Metabolism studies can also be carried out by incubating BMEC with drugs and subsequently determining the metabolite(s) generated (22). Interaction studies can be performed in order to characterize carrier- and receptor-mediated transport (23), while the application of endotoxins allows the study of the up- or down-regulation of adhesion molecules at the BBB (24). Studies to visualize the transport route of fluorescent drugs or probes through the endothelial monolayers have been performed with confocal laser scanning microscopy (CLSM) (25). This technique is particularly useful for the assessment of changes in the intracellular concentration of Ca^{2+} and Na^{+} ions, pH, and the cytoskeleton induced by absorption enhancers and hyper- or hypo-osmotic solutions (26, 27).

The results obtained with this *in vitro* model can be influenced by various factors that change the culture conditions and, therefore, change the expression of surface molecules and receptors. Thus the presented model has

Table 1. γ-GTP activity of BMEC cultured with astrocytes or in ACM

Culture time (days)	γ-GTP activity (units/mg of cell protein) Cultured alone in normal medium	Co-cultured with astrocytes	Cultured with ACM
10	5.5 ± 0.81	—	—
15	4.9 ± 0.75	9.5 ± 1.83	6.3 ± 1.18

its limitations, and thorough characterization of the system is important before performing and interpreting experiments. Nevertheless, this *in vitro* BBB model is an important tool for developing and studying various concepts and strategies for drug transport to the brain. Ultimately the data obtained with this model should be verified *in vivo*.

References

1. Bradbury, M. (ed.) (1979). *The concept of a blood–brain barrier*. John Wiley & Sons, Chichester, UK.
2. Reed, D. J. (1980). In *Antiepileptic drugs: mechanisms of action* (ed. G. H. Glaser, J. K. Pentry, and D. M. Woodbury), pp. 199–205. Raven Press, NY.
3. Janzer, R. C. and Raff, M. C. (1987). *Nature*, **325**, 253.
4. Arthur, F. E., Shivers, R. R., and Bowman, P. D. (1987). *Dev. Brain Res.*, **36**, 155.
5. Tao Cheng, J. H. and Brightman, M. W. (1988). *Int. J. Dev. Neurosci.*, **6**, 25.
6. Minakawa, T., Bready, J., Berliner, J., Fisher, M., and Cancilla, P. A. (1991). *Lab. Invest.*, **65**, 32.
7. Pardridge, W. M. (ed.) (1991). *Peptide drug delivery to the brain*. Raven Press, NY.
8. Weindl, A. (1973). In *Frontiers in neuroendocrinology* (ed. W. F. Ganong and L. Martini), pp. 3–32. Oxford University Press, NY.
9. Van Bree, J. B. M. M., De Boer, A. G., Danhof M., and Breimer, D. D. (1992). *Pharm. Weekblad Sci. Ed.*, **14**, 305.
10. Méresse, S., Dehouck, M-P., Delorme, P., Bensaïd, M., Tauber, J-P., Delbart, C., Fruchart J-C., and Cecchelli, R. (1989). *J. Neurochem.*, **53**, 1363.
11. Rubin, L. L., Hall, D. E., Parter, S., Barbu, K., Cannon, C., Horner, H. C., Janatpour, M., Liaw, C. W., Manning, K., Morales, J., Tanner, L. I., Tomaselli K. J., and Bard, F. (1991). *J. Cell Biol.*, **115**, 1725.
12. Van Bree, J. B. M. M., de Boer, A. G., Danhof, M., and Breimer, D. D. (1992). *Pharm. Weekblad Sci. Ed.*, **14**, 338.
13. Audus, K. L. and Borchardt, R. T. (1986). *Pharm. Res.*, **3**, 81.
14. Tio, S., Deenen, M., and Marani, E. (1990). *Eur. J. Morphol.*, **28**, 289.
15. Rubin, L. L., Hall, D. E., Parter, S., Barbu, K., Cannon, C., Horner, H. C., Janatpour, M., Liaw, C. W., Manning, K., Morales, J., Tanner, L. I., Tomaselli, K. J., and Bard, F. (1991). *J. Cell Biol.*, **115**, 1725.
16. McCarthy, K. D. and De Vellis, J. (1980). *J. Cell Biol.*, **85**, 890.

17. Meresse, S., Dehouck, M-P., Delorme, P., Bensaid, M., Tauber, J.-P., Delbart, C., Fruchart, J.-C., and Cechelli, R. (1989). *J. Neurochem.*, **53**, 1363.
18. Hurni, M. A., Noach, A. B. J., Blom-Roosemalen, M. C. M., de Boer, A. G., Nagelkerke, J. F., and Breimer, D. D. (1993). *J. Pharmacol. Exp. Ther.*, **267**, 942.
19. Risau, W., Dingler, A., Albrecht, U., Dehouck, M-P., and Cechelli, R. (1992). *J. Neurochem.*, **58**, 667.
20. de Vries, H. E., Kuiper, J., de Boer, A. G., van Berkel, Th. J. C., and Breimer, D. D. (1992). *J. Neurochem.*, **61**, 1813.
21. van Bree, J. B. M. M., de Boer, A. G., Danhof, M., Ginsel, L. A., and Breimer, D. D. (1988). *J. Pharmacol. Exp. Ther.*, **247**, 1233.
22. van Bree, J. B. M. M., de Boer, A. G., Verhoef, J. C., Danhof, M., and Breimer, D. D. (1989). *J. Pharmacol. Exp. Ther.*, **249**, 1836.
23. de Vries, H. E., Kuiper, J., de Boer, A. G., van Berkel, Th. J. C., and Breimer, D. D. (1993). *J. Neurochem.*, **61**, 1813.
24. de Vries, H. E., Moor, Anne C. E., Blom-Roosemalen, Margret C. M., de Boer, A. G., Breimer, D. D., van Berkel, Th. J. C., and Kuiper, J. (1994). *J. Neuroimmunology*, **52**, 1.
25. Jaehde, U., Masereeuw, R., de Boer, A. G., Fricker, G., Nagelkerke, J. F., Vonderscher, J., and Breimer, D. D. (1994). *Pharm. Res.*, **11**, 442.
26. Shotton, D. (1989). *J. Cell Sci.*, **94**, 175.
27. Rojanasakul, Y., Paddock, S. W., and Robinson, J. R. (1990). *Int. J. Pharm.*, **61**, 163.

Index

ACE, *see* angiotensin converting enzyme
actin 99, 118–20
adenine 184
adenylate cyclase 39
alkaline phosphatase
  expression of 39, 209
  labelled antibodies 24, 25–6, 34
aminopeptidases 50
amphotericin B 91–2
angiotensin converting enzyme 101, 103, 209
antibodies
  to basement membrane components 194, 196
  to cytokeratins 18, 22–3
  to endothelial cells 100–2
  to inhibin 173
  to junctional proteins 50, 119–21, 131–2
  to receptor proteins 50, 59–61, 151
astrocytes 206, 208–10

basement membrane 7, 8, 9, 14, 194, 195
bisBenzimide 119, 121, 171
blood–brain barrier 205–10
blood–testis barrier 159–77
brush border enzymes 111
bullous pemphigoid antigen 14, 194

Caco-2 cells 12, 52, 111–31
cadherin 50, 101–2
cadmium chloride 170–5
CD31 101–2
CD44 50
cell density 119
cell number 171–2, 174
cell polarity 6–7, 37–66
cell viability 118–19, 170–2, 174–6
ceramides 14
cholera toxin 184, 190
collagen 9, 12, 26, 113, 194
collagenase 141–3, 164–8, 206–7
collagen-fibroblast lattice, *see* dermal equivalent
confocal scanning laser microscope 50, 51, 130–1, 209
cornification 14, 194, 197
corticosterone 127
cryopreservation
  of endothelial cells 103–5, 206–7
  of 3T3 fibroblasts 185
  of keratinocytes 186
  of MDCK cells 41, 42
cytochrome P450
  in Caco-2 cells 111
  chlorzoxazone 6-hydroxylase activity 75–6
  dextromethorphan demethylase activity 74–5, 83
  ethoxyresorufin dealkylase activity 70, 71–2, 83–4
  lauric acid hydroxylase activity 78–9, 83
  measurement of total 70–1
  mephenytoin hydroxylase activity 73–4, 83
  nifedipin oxidase activity 77, 83
  pentoxyresorufin dealkylase activity 71–2, 83
  phenacetin deethylase activity 72–3, 83–4
cytokeratins
  antibodies to 18, 22–3
  classification of 6, 18–19
  identification/localization of
    electrophoresis 27–32
    immunoblotting 33–5
    immunohistochemistry 18–26
    Western blotting 32–3
  K1/K10 6, 14, 194, 196
  K3/K12 6
  K4/K13 6, 26
  K5/K14 6, 14, 26, 30
  K6/K16 6, 14, 30
  K8/K18 30
  K19 30
cytoskeleton 38

de-epidermized dermis 179–80, 187–9, 193, 197
2-deoxyglucose 124–6
dermal equivalent 26, 179–83, 187–92
desmosome
  belt 7, 8, 38
  hemi- 7, 8, 14, 38, 195
  labelling 50
  spot 7, 8, 14, 38, 54, 195
dexamethasone 60–2
dextran 130–1
DNA content 171–2
Dynabeads 95, 96, 102

Index

ECGF, *see* endothelial cell growth factor
EGF, *see* epidermal growth factor
EN-4 antigen 101–2
endocytosis 55, 59, 61
endothelia 99–109
  microvascular 2–3, 87, 96–9, 205–10
  umbilical vein 87, 92–4
endothelial cell growth factor 88–9, 99
endotoxin 103, 209
enterocyte 7, 111
epidermal growth factor 138, 165, 168, 184, 190
epithelia
  buccal 26, 30
  corneal 6
  duodenal 2–3
  endometrial 135–56
  intestinal 6–7, 111–31
  mammary 135–6
  oesophageal 6
  pancreatic 136
  renal 2–3, 6, 39–63
  skin (epidermis) 3, 6, 14, 179–98
  thyroid 136
  tracheal 2–3, 136
  ureter 3
  uterine 135
  vaginal 135
estrogens, *see* oestrogens

FACS 95, 96, 98
feeder layer 9, 179, 183–7
fenestrae 205
fibroblast growth factor 97, 99
fibronectin 9, 12, 44, 89–90, 140, 194, 206–7
filaggrin 14, 194, 196
filters
  polycarbonate 113
  types available 43–4
fluorescein isothiocyanate 23–4, 50, 100–2, 130–1
fluorescence activated cell sorter, *see* FACS
flux rate constants 122, 125–6
follicle stimulating hormone 170, 174
forskolin 105, 107
Forssmann antigen 39

gap junction 7, 8, 38
gelatin 89–90
germ cells 159, 161, 163–4, 168, 174
glial fibrillary acidic protein 208
glucuronidation
  in Caco-2 cells 111
  of 4-methylumbelliferone 79
  of paracetamol 80, 83–4
γ-glutamyl transpeptidase 103, 209, 210
glutathione-*S*-transferase
  in Caco-2 cells 111
  determination of activity 81–2, 83
Golgi apparatus 52–4, 195

hair follicles 179, 186, 191–3
hepatocytes
  culture of 67–9
  membrane trafficking of proteins 38
  xenobiotic metabolism 67, 69–84
horseradish peroxidase
  assessment of monolayer integrity 46–7, 105, 107–8
  labelled antibodies 24, 26, 34–5, 53–5, 151
hydrocortisone 127, 184, 190

immunogold 55
inhibin secretion 170–1, 172–6
insulin 165, 184
interleukin-1 103, 198
interleukin-6 103
involucrin 14, 194, 196

keratin, *see* cytokeratins
keratinocytes
  culture of 179–98
  xenobiotic metabolism 14, 67, 70, 84
keratohyaline granules 14, 194, 197

laminin 9, 12, 38, 44, 194
lanosterol 14
Leydig cells 159, 161, 164, 177
low density lipoproteins, DiI-acetylated 90–1, 98, 101, 103, 209

macula adherens, *see* desmosome, belt
mannitol 115–17, 127
Matrigel 12
  coating of filters 140
  endometrial cell culture 136–7, 143, 145–56
  hepatocyte culture 68
  Sertoli cell culture 160–3, 165, 170
MDCK cells 12, 39–66
membrane coating granules 14, 194
membrane proteins
  biotinylation 56–61

Index

radio-iodination 56, 59–62
trafficking 38, 59–63
methionine 138, 148–50
metoprolol 127
microcarriers 106–8
microsomes 69–70
microvilli 6–7, 39, 50, 52–3, 119, 160
mitomycin C 184, 187
monoamine oxidase 103
monolayer integrity 111–12
assessment using HRP 46–7, 105, 107–8
assessment using inulin 162, 169
assessment using mannitol 115–17
assessment using TER 47, 114–15
MTT assay 171, 172, 175
mycoplasma 91, 113

N-acetylation 82–4
Na/K-ATPase 41, 123–4, 126

oestrogens 135, 149, 151–3
ouabain 125–6

PAL-E antigen 101–2
pericytes 209
peritubular myoid cells 161, 164, 168, 169
permeability coefficient
apparent 116–17, 122–3, 127–31
cell 127–31
filter 129–31
phalloidin 119–20
phenol red 137
phthalate esters 159, 174–6
pIg-R antigen 59–62
pinocytotic vesicle 205
polarity, *see* cell polarity
poly-D-lysine 140
progesterone 135, 149–56
propidium iodide 118–19
propranolol 127
protein secretion 59, 135–6, 147–56, 159, 161–2, 170–6
proteoglycans 9, 38

retinoic acid 30, 165, 188, 197
retinol acetate 138
ruthenium red 120

salicylic acid 127
secretion, *see* protein secretion
selectin 100, 103
Sertoli cells 159–77
sodium azide 124–6
sphingolipids 14
stromal cells 135–7, 142, 153–6
sulfasalazine 127
sulfation
in Caco-2 cells 111
of paracetamol 80, 83

TER, *see* transepithelial electrical resistance
testosterone 127, 129, 170
thrombin 105, 107
Thy-1 antigen 60–2
tight junction 7, 8, 37–8
ATP depletion, effect of 124
brain endothelia 205, 209
Caco-2 cells 111
immunostaining of 39, 50, 119–21
MDCK cells 39, 50, 53–4
Sertoli cells 159–62, 169–71, 176
toluidine blue 45
transcytosis 59, 61, 122, 131
transepithelial electrical resistance
of brain endothelial cell monolayers 208, 209
of Caco-2 monolayers 115
and cytotoxicity studies 174–6
of endometrial cell monolayers 147
of MDCK monolayers 39, 40, 44–5, 47
measurement of 114–5, 170
of Sertoli cell monolayers 162, 169–71, 174–6
transferrin
addition to media 138, 165, 184
internalization 55–7
receptors 50
transglutaminase 14, 194, 196
transport processes 111–31, 209–10
triglycerides 14
tri-iodothyronine 184
tri-*o*-cresyl phosphate 177
tumour necrosis factor 103

Ulex europaeus lectin-1 101–2
unstirred water layer 126–31
urokinase-type plasminogen activator 103

vimentin 26
vitamin A, *see* retinoic acid
vitamin D 197

vitamin E 165
von Willebrand factor 100–1, 209

warfarin 127
Weibel–Palade bodies 100

ZO-1 tight junction protein 39, 50, 119–21, 131–2
zonula adherens, *see* desmosome, belt
zonula occludens, *see* tight junction